Grade 7
World Cultures and Geography

Nancy F. Artis

Mary Enda Costello

Robert Miltner

Contributors

Roberta J. Leach

Karen Martin Tryda

Curriculum Unit

Nancy F. Artis earned a B.S. degree from Savannah State College and M.S. degrees from both City College and Hunter College.

Mary Enda Costello earned her M.A. at the University of Notre Dame, Indiana. As a secondary teacher and adviser, she authored Center for Learning curriculum units in the language arts and novel/drama series, as well as social studies.

Robert Miltner, coordinator of developmental education at Kent State University, earned his M.Ed. at John Carroll University, Cleveland, Ohio. The former English teacher and department chairman is a nationally published poet and the author of novel/drama curriculum units published by The Center for Learning. He is also coauthor of the Center's *World History, Book 2* curriculum unit.

The Publishing Team

Rose Schaffer, M.A., President/Chief Executive Officer
Bernadette Vetter, M.A., Vice President
Lora Murphy, M.A., Vice President, Social Studies Division
Amy Richards, M.A., Editor

Cover Design

Krina K. Walsh, B.S.I.D.

List of credits found on Acknowledgments page beginning on 240.

ISBN 1-56077-489-4

Contents

Introduction

The New Electronic interdependence recreates the world in the image of a global village.
—Marshall McLuhan

Geography has traditionally studied the landmasses of the earth and the individual characteristics of each landmass: temperatures, plants, climate, resources, distinctive features, and inhabitants. This represents the physical geography curriculum with which we are all familiar. However, a study of geography should also probe the cultures of those inhabitants and interpret these cultures in relation to the geographic areas that support them. The five fundamental or basic themes espoused by geography educators in *Guidelines for Geographic Education* reflect this approach.

Culture creates discernible patterns that show how people work and play, what they believe in and what they value, how they define arts, crafts, and technology. Essentially, culture shows how a group of people view themselves both as distinct from and in relation to others. A culture is neither superior nor inferior—it is simply different and unique.

An understanding of the multiplicity of world cultures is the key to unlocking the shackles of misunderstanding. Understanding cultures helps a person learn the nature of humanity, in all of its diversity, that is accessible to us today in the global village of human beings on this planet.

Part 1
Studying Our Global Village

We live today in an age of instant information, of integrated circuits of knowledge, in an age of microchip and telecommunication interaction. We live in a global village where we share gossip and news across the backyards of the world.

Yet, while we are one world, we are different people, as different as the birds in the air, the fish in the sea, or the flowers that bloom across the landmasses of the earth; each one different, each one unique, each one with its own special significance.

In order to learn more about who we are, both as a people and as members of a culture, we examine our similarities and differences.

If we can learn more about where we live by studying geography, and we can learn more about how we live by studying culture, then perhaps we can learn how to live together in peace and understanding in this global village of humanity.

Lesson 1
The Five Geographic Themes

Objectives

- To identify the five geographic themes
- To use the five geographic themes to analyze the effect of human and natural intervention upon the earth's surface

Notes to the Teacher

This lesson develops the five geographic themes as outlined by the National Geographic Society. It is important for students to integrate the geographic themes into their study of geography, history, economics, and ecology because the themes help students understand the meaning of the world around them.

This lesson takes students through the five geographic themes so that they can identify the themes and use them to analyze the relationship among systems and processes on earth. Using the five themes, students solve a problem related to a current situation. They make a decision about the problem by analyzing the geographic information they have gathered and predicting the effects of their decision on the earth.

Procedure

1. Show pictures of various parts of the United States where humans have settled and changed the environment. (Arid areas where many people have settled—Arizona, New Mexico, California, Colorado; Arctic areas—Alaska; wet, lowlands—Florida; highly populated areas such as New York City, Boston; and other locations.)

2. Brainstorm how people interact with their environment. Discuss how people build dams to divert water to dry regions, use fill dirt to reclaim swampy lowlands for building, scar the Arctic soils to build oil pipelines, build concrete highways in cities along with high office towers and apartment buildings over soil that could be used for agriculture, etc. Have students explain how these activities change the earth and what these effects have to do with the future of the earth and its population of people and animals.

3. Use **Handout 1** to introduce the five geographic themes. Discuss each of the themes, using the examples provided. Select some other examples for students to locate in a student atlas. Have students give *Location*, absolute and relative, and tell what they know about its *Place*, *Human-Environment interactions* there, and examples of *Movement* and *Regions*.

   Example: *Japan*

   Location
   Absolute—*38° N, 138° E*

   Relative—*East of Korea, Southeast of China, West of California, in the Pacific Ocean*

   Place
   Physical characteristics—*rugged, mountainous, and archipelago*

   Human characteristics—*large population, little diversity of culture, speak Japanese, willing to work cooperatively/dislike competition*

   Human-Environment interactions
   Only a small part of the country is level and easily farmed, so inhabitants terrace the hillsides to produce food. They have few natural resources so they import oil and other resources to use in their successful manufacturing processes.

   Movement
   Many Western ideas including baseball, rock music, Western clothing styles, and Christian religions have entered Japan since World War II. Japanese culture has been influenced greatly by this movement of Western ideas.

   Region
   Japan is part of Eastern Asia. It is an archipelago—volcanic mountaintops protruding from the waters of the Pacific Ocean. People were farmers and fishermen until manufacturing became such an important part of their lifestyle. This region produces most of the world's electronic products and cars and markets them in many countries throughout the

world. Japan's culture, until World War II, remained the same over the centuries when westerners began to be an influence.

4. Distribute **Handout 2** and discuss the problem, making sure students understand the task.

5. On a map, locate Santa Fe. Student atlases are a useful resource. Find out about the place and the region. Do the same for Lake Michigan; Chicago, Illinois; and Toronto, Ontario, Canada. Distribute **Handout 3** and the map, and have students circle these places.

Suggested Responses; Handout 3

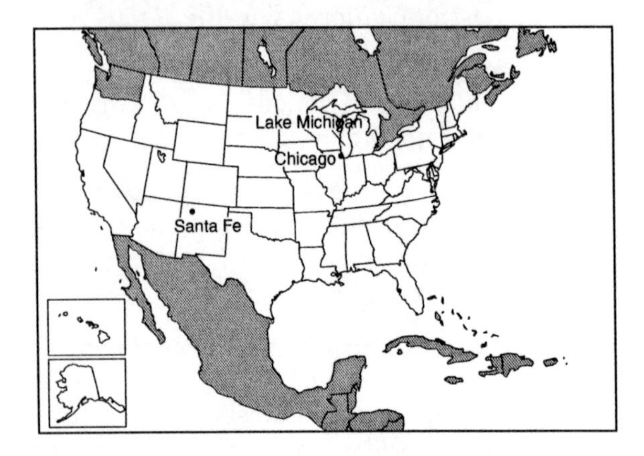

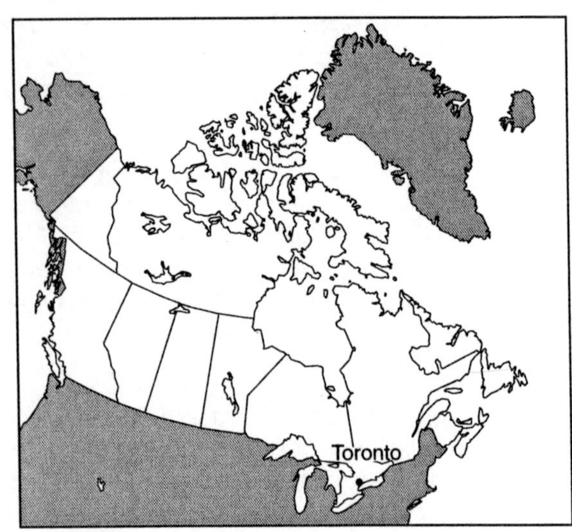

6. Divide the class into small groups of four or five students per group. Have them use encyclopedias and/or library books about states to find out about the Human-Environment interaction and movement in Santa Fe. (For example: What kind of culture is there? What are the social and economic backgrounds of the residents? What kinds of industry/business do they have? Why might more people want to live there?)

7. Have one half of the groups prepare arguments that support the attempt to get water from Lake Michigan with the other half of the groups preparing arguments against getting water that way, using their knowledge of the geographic themes to support their arguments. Students should be able to discuss the effects of their plan upon Santa Fe, Michigan, and Canada. (Political and international considerations should be listed in students' rationale since losing water from the Great Lakes would affect not only all the Great Lakes Region but Canada as well.) Have students use **Handouts 2** and **3** to prepare their arguments.

8. Have students present their arguments to the class and, at the conclusion of all the groups' presentations, have the class vote on the issue: Is it wise to move the resource, water, from the Great Lakes to an arid state in the southwest? Would you support such an action—yes or no? Why or why not?

Enrichment/Extension

1. Organize a formal debate on the following question:

   Resolved: We should move water from its original source to drier regions to support a larger population base. Other issues can be debated.

2. Use the geographic themes to look at other places, charting them out to examine other issues that are particular to them.

Name________________
Date________________

Geography's Five Fundamental Themes

Examine the following:

Location
This tells "where" and "why" something is found on the earth. There are two ways to discuss the location of a site:

- *Absolute location*—This is the *exact* position of a place on the earth's surface. Latitude and longitude are used to name its position.
- *Relative location*—This is where the location of a site is in relation to other sites that influence us. Knowing what else is near a site and how those areas influence a location explain "why" things are located in particular spots.

Example

- *Absolute location*—42° N, 82° W, Cleveland, Ohio
- *Relative location*—Cleveland is located on the southern shore of Lake Erie and at the mouth of the Cuyahoga River. Water transportation was the only way to transport heavy goods in and out of the area during its settlement period. Later, its location on Lake Erie made it a hub for national and international trade.

Place
Every spot on the earth can be identified by its physical and human characteristics. If the physical and human characteristics of a place are known, then one can understand better the relationship between people and their environment. Place tells what it is like at that spot.

- *Physical characteristics*—The landforms, bodies of water, climate, soils, natural vegetation, and animal life that form the natural environment are the physical characteristics of a place.
- *Human characteristics*—The population of a place, architectural styles, kinds of economic activities, types of recreational activities, languages, religions, social and political organizations, etc., make up the human ideas and the activities that give shape or character to a place.

Example

Cleveland, Ohio, sits on a lake plain, but there are deep valleys created by the same glacier that formed Lake Erie. The city is surrounded by forests that have been preserved as park lands. The downtown area has tall buildings with the Key Corp Tower on Public Square being the tallest building in the state of Ohio. There is a great variety of housing in and around the city, varying from historic Greek Revival and Victorian architecture to ultra-modern structures. Many new buildings like the Rock and Roll Hall of Fame have been built to interest visitors in the city.

Name________________
Date________________

Human-Environment interactions
Because there are good things and bad things about a place, humans have learned to change the environment to meet their needs based upon their cultural, social, and/or economic backgrounds. Their level of technology affects people's ability to effect change. This helps us understand the relationship between people and the environment.

Example

The Cuyahoga River that divides the city of Cleveland in half has provided the city with not only a way to transport goods in and out of the city, but people have used its water to supply the industries that line its banks. These same industries eventually polluted the water, and people had to change their ways and clean the water so that the fish would stop dying and the water could be used. In Cleveland's early days, people settled along its shores but soon learned that the mosquitoes that shared the space caused them to be ill. They moved to higher ground to build their homes, and used the river for industry.

Movement
People in one place communicate and trade with people from other places. Transportation of people, ideas, and goods among places demonstrates global interdependence. This explains "how" and "why" places are related to one another.

Example

In the 1940s, many African Americans moved to Cleveland because there was work for them in the industries that were booming in the area. They came from the South where prejudice and poverty kept them from employment. Cleveland still has two major automobile firms (Ford and Chevrolet) that manufacture car parts. These goods are sent all over the nation as well as to other countries.

Regions
Considered to be the basic units when studying geography, regions are identified by a characteristic that is common within a given area. To define a region, human characteristics (farming, mining, manufacturing), physical characteristics (mountains, plains, etc.), or both may be used.

Example

The Region where Cleveland is located is usually considered part of the Midwest. When people talk about this region, they mention the prosperous farming, manufacturing, and business activities that take place. Geographically, the major portion of this region was affected by the last glacier leaving behind lake plains in the northern sections and hilly regions in the southern sections where the glacier stopped.[1]

[1]Adapted from Joint Committee on Geographic Education of the National Council for Geographic Education and the Association of American Geographers, *Guidelines for Geographic Education* (Washington D.C.: Association of American Geographers, 1984), 3–8.

Name________________
Date________________

Five Geographic Themes

Read the following problem. Use reference material from the library about Santa Fe, New Mexico, to complete the chart.

The residents of Santa Fe, New Mexico, want their population to grow, so they must find an additional source of water. Their location in a desert region makes this a serious issue. Mr. Alvarez, the local chairman of the city resources committee, has discovered that the only alternative is to have water piped in from somewhere other than the Colorado River (because too many states are already using its water). Could Lake Michigan's water be piped to Santa Fe?

Location
Absolute

Relative

Place

Human-Environment interaction

Movement

Regions

Name________________
Date________________

The Five Geographic Themes

Locate and circle the following places: Santa Fe, New Mexico; Lake Michigan; Chicago, Illinois; and Toronto, Canada.

The United States

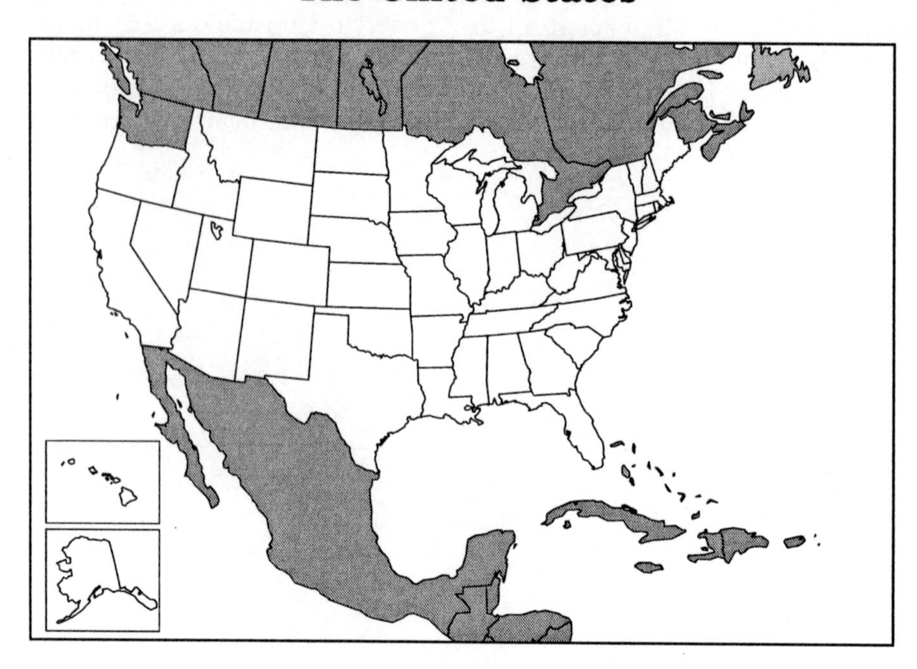

Canada

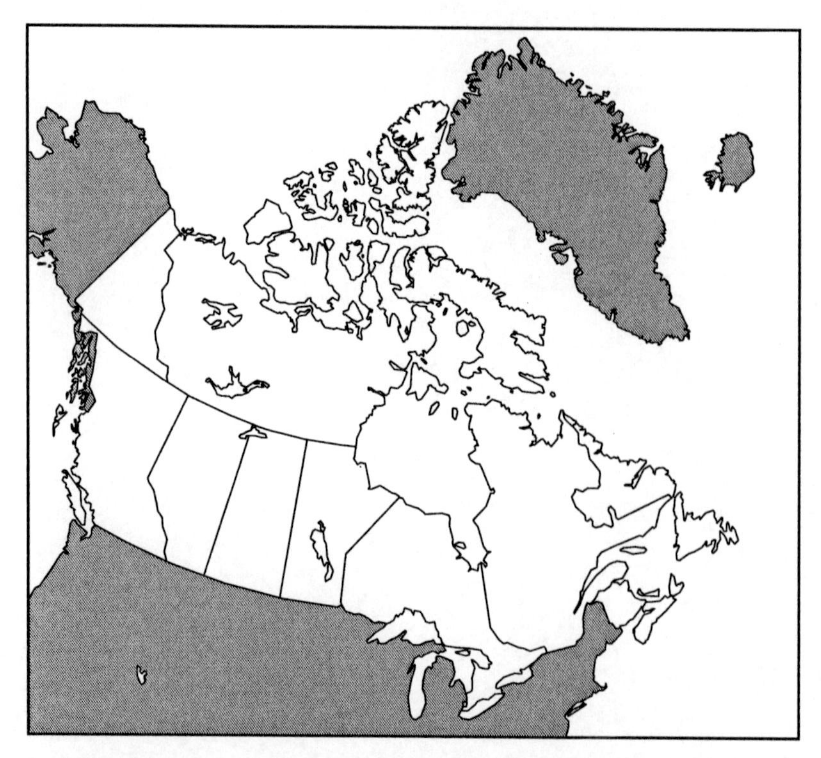

Lesson 2
Geography Is . . .

Objectives

- To develop a working definition of geography
- To become familiar with basic concepts and terminology necessary for work with globes and maps

Notes to the Teacher

This lesson gives students a basic understanding of geography as a study of the relationship between people, places, and regions of the world. A study of geography enhances students' understanding of the interrelationships between human society and the physical world and helps to explain historical events that they will study in the future. Make students aware that they will be studying many peoples who have the same basic needs that they have. Pictures, readings, and maps help to bring to life faraway places that students may not have seen.

Have ready a supply of blue balloons; markers; crayons or paints; scissors; paste or glue; and string or yarn.

Students work with desk maps to gain familiarity with key terminology related to the study of geography. To demonstrate their grasp of key concepts, students create their own simple globes from blue balloons.

Procedure

1. Help students develop a working definition of geography (*e.g.*, a study of the relationship between people, places, and regions of the world). Ask students what they expect to learn in a class focusing on the relationship between people, places, and regions of the world. Record their responses on the chalkboard.

   Suggested Responses:

   How different regions of the world are related to one another

   How the environment influences ways of living or working in different regions

   Differences in lifestyles in different regions of the world

   How to locate places on maps and globes

2. Write the word "geography" on the board and explain that the term came from the ancient Greeks. (A Greek scholar originated the word "geography" by combining two common Greek words: ge—the earth—and graphein—to write.)

3. Write the following words on the chalkboard. Pronounce each word and explain how it helps in locating places on a map and/or globe.

equator	South Pole
parallels of latitude	continent(s)
latitude	hemisphere(s)
North Pole	meridian of longitude
longitude	prime meridian

4. Distribute **Handout 4** and complete the exercise with the class as a whole. A transparency of the handout is useful. Be sure to point out the equator as the reference point in determining latitude and the prime meridian as the reference point in determining longitude.

   Suggested Responses:

   1. *Asia*
   2. *Africa*
   3. *North America*
   4. *South America*
   5. *Antarctica*
   6. *Europe*
   7. *Australia*
   8. *compass rose*
   9. *equator*
   10. *parallel of latitude*

11. *Northern Hemisphere*

12. *Southern Hemisphere*

13. *Eastern Hemisphere*

14. *prime meridian*

15. *meridian of longitude*

16. *North Pole*

17. *Western Hemisphere*

18. *South Pole*

5. Distribute **Handout 5** and have students complete it individually as a check of their understanding of key terms.

Suggested Responses:

1. *equator*
2. *hemispheres*
3. *continents*
4. *hemisphere; hemisphere*
5. *longitude*
6. *prime meridian*
7. *latitude*
8. *North Pole*
9. *South Pole*
10. *compass rose*

6. Distribute **Handout 6** and have students demonstrate their command of the key concepts by making the suggested globe. Depending on the sophistication of students, you may wish to introduce the concept of a physical map and suggest that students try to color the continents appropriately. For less advanced students you may choose to have them simply color the continents green or brown. Use completed globes for a classroom display.

Enrichment/Extension

Give an oral presentation which demonstrates to your classmates how your completed globe illustrates the vocabulary introduced in this lesson.

Name____________________
Date____________________

Building Your Geographical Vocabulary

On the maps below, locate and write in the names of each of the following items: prime meridian, equator, North Pole, South Pole, North America, South America, Africa, Europe, Asia, Australia, Antarctica, Northern Hemisphere, Southern Hemisphere, Eastern Hemisphere, Western Hemisphere, compass rose, parallel of latitude, meridian of longitude.

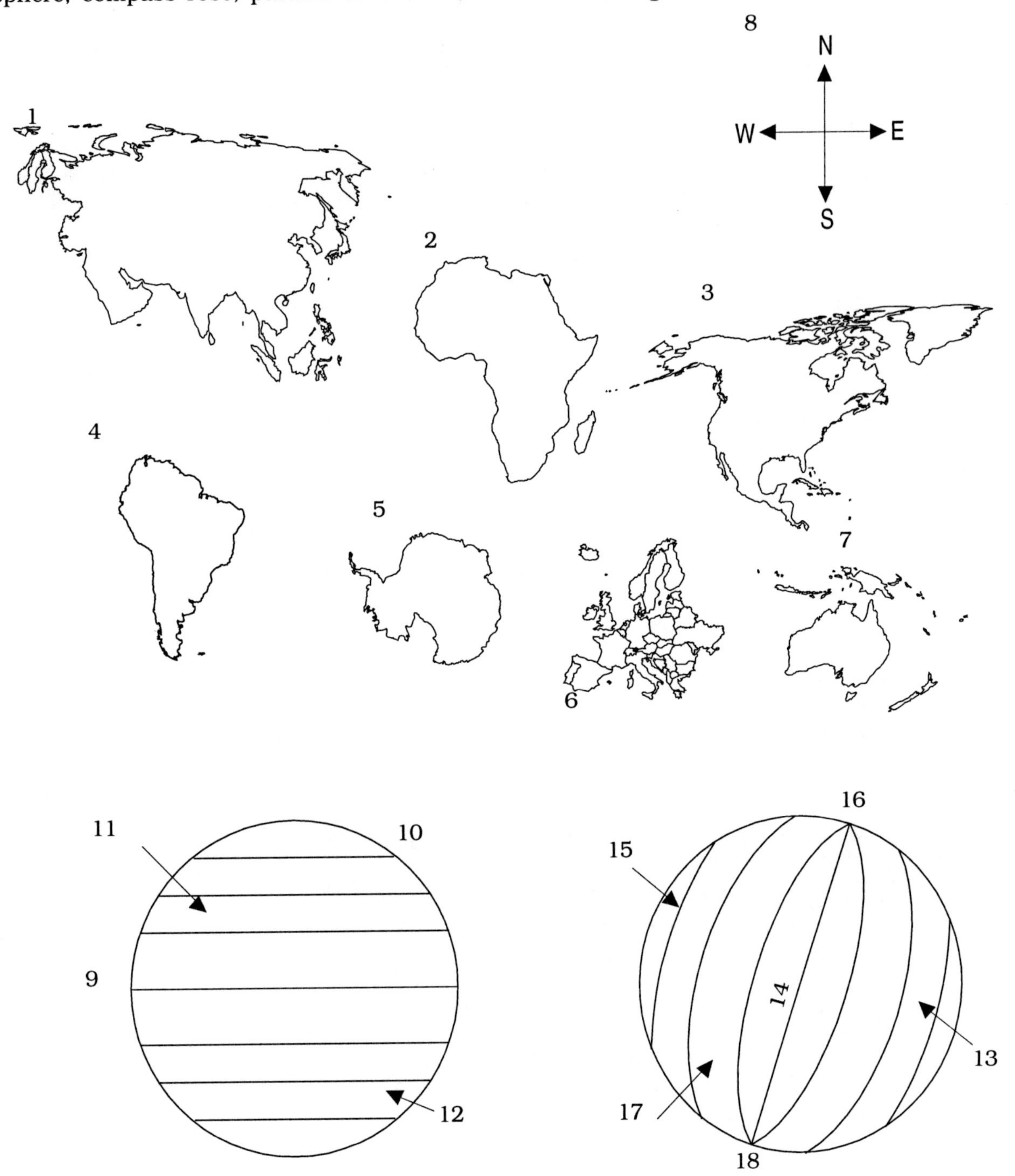

Name__________________
Date___________________

Globe Skills

To do the exercise below, use the previous handout which you completed.

1. The imaginary line that represents 0° of latitude is called the __________________________.

2. The earth is divided into two equal parts called _____________________________________.

3. The seven __________________________ are Europe, Asia, Africa, Antarctica, Australia, North America, and South America.

4. There is a northern ______________________________ and a southern _____________________.

5. ____________________________________ measures distance in degrees east and west.

6. The ________________________________ is the imaginary line that represents 0° of longitude.

7. ______________________ measures distance in degrees north and south of the equator.

8. The northernmost polar region of the Northern Hemisphere is called the _______________.

9. The southernmost polar region of the Southern Hemisphere is called the ______________.

10. A ___________________ indicates direction on a map or globe.

Name________________
Date________________

Creating a Globe

Demonstrate your understanding of the vocabulary and concepts you have learned by creating your own globe using the following steps.

1. Color the continents on the following page to indicate land forms.
2. Write in the names of the continents.
3. Cut out the continents.
4. Blow up a blue balloon.
5. Use a marker to create a compass rose on your globe.
6. Use a marker to locate the North Pole and South Pole.
7. Paste the continents in appropriate places on your globe.
8. Use a piece of string tied loosely around the center of your globe to serve as a guide in drawing in the equator. Use a marker for this purpose.
9. Use the marker and string to draw in the prime meridian. It goes through Greenwich, England, a suburb of London.
10. Complete your globe by labeling these four oceans: Atlantic, Pacific, Indian, and Arctic.

Name________________

Date________________

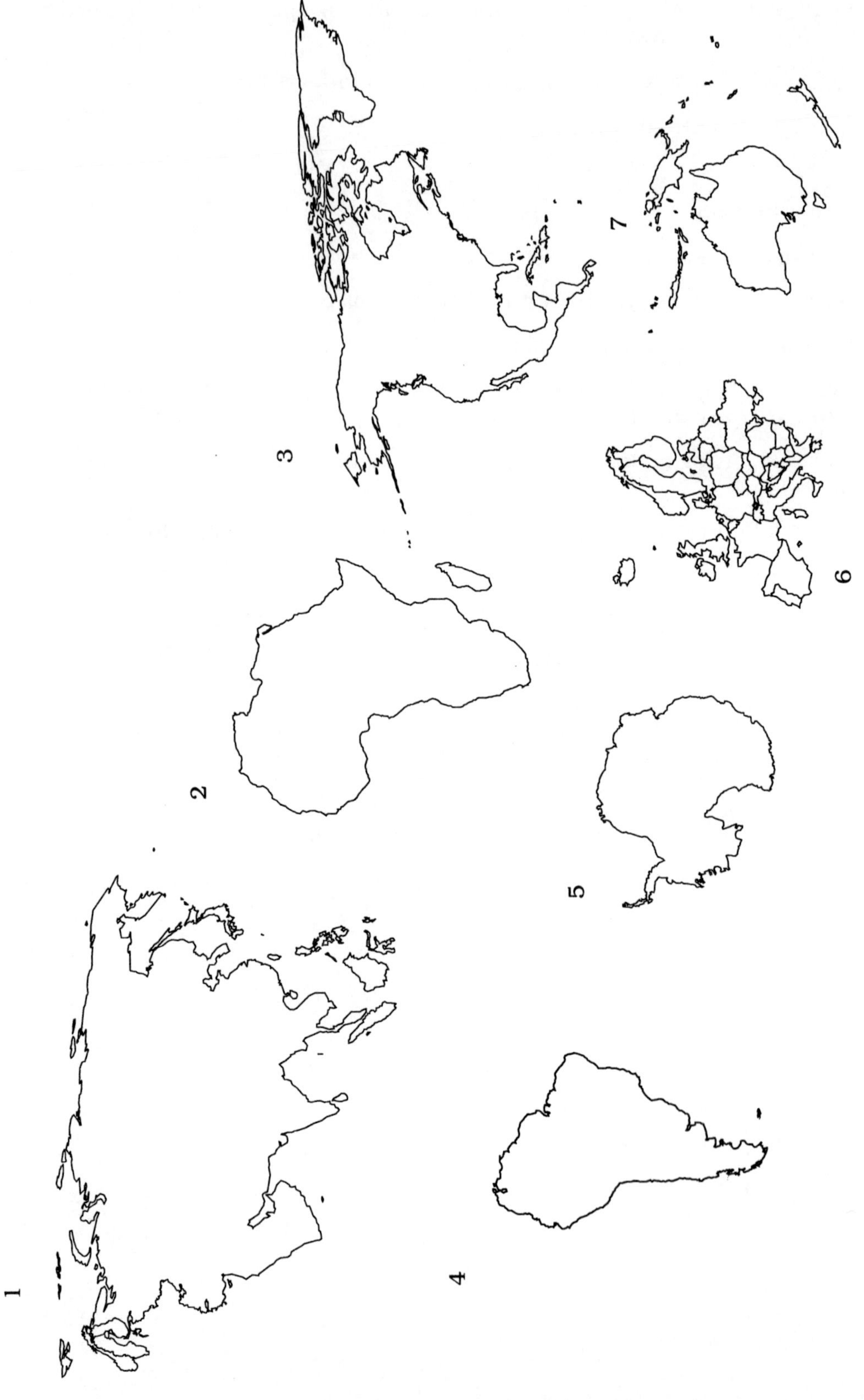

Lesson 3
Reading Maps

Objective

- To review and practice map reading skills

Notes to the Teacher

This lesson extends the previous lesson and provides practice in skills needed in map reading. The exercises included are intended to supplement maps in the text.

Students being introduced to new map reading skills often benefit from tactile experiences. Thus, preparation of classroom manipulatives pays big dividends in the long run. Enlarged maps that have been either laminated or mounted on cardboard are useful. The grid in **Handout 7** and the Mercator projection world map in **Handout 9** are good examples. Initial preparation of learning devices requires an investment in time and money, but these teaching aids can be saved for use in future years.

Students practice using grids, measuring distances, finding latitude and longitude, and reading topographical maps. At the conclusion of the exercise, students should have confidence in their command of a variety of skills needed in using maps and globes.

Procedure

1. Begin by reviewing with students the definition of geography and the terms presented in the previous lesson.

   Suggested Responses:

   Definition—*a study of the relationship between people, places, and regions of the world*

   Terms—*equator, parallels of latitude, latitude, longitude, North Pole, South Pole, continent(s), hemisphere(s), meridian of longitude, prime meridian*

2. Distribute **Handout 7** and display your grid map of **Handout 7** if you made one. Give students time to complete the activity. Have students share their responses to determine whether or not more time is required on this particular skill.

   Suggested Responses:

   1. Morristown, Paterson

   2. A6

   3. H1, H2

   4. E2

   5. B5, C5

3. Distribute **Handout 8** on map scales. Allow time for students to complete the exercises and have students share responses to determine whether or not more time is required on this particular skill.

   Suggested Responses:

   1. 180 miles

   2. 300 miles

   3. 140 miles

   4. 50 miles

   5. 200 miles

4. Distribute **Handout 9** on reading latitude and longitude. Explain the inaccuracies of a Mercator map. Point out the difficulty of putting a round globe on a flat map. Explain that this map is useful in teaching latitude and longitude and that it is reasonably accurate in the areas near the equator. Have students tie a band of yarn around the world map horizontally and another vertically to create a useful device for finding latitude and longitude. Allow students time to answer the questions. Have students share their responses so that they get immediate feedback on the exercise to determine if more time is needed on this particular skill.

   Suggested Responses:

   Part A.

   1. South America

   2. Atlantic Ocean

   3. Africa

   4. Pacific Ocean

   5. Indian Ocean

6. *Europe*
7. *Atlantic Ocean*
8. *Australia*
9. *Asia*
10. *Canada*
11. *Europe*
12. *Pacific Ocean*
13. *Pacific Ocean*
14. *The same number of degrees applies to different places on the globe.*

Part B.

1. *34° south latitude; 18° east longitude*
2. *36° south latitude; 58° west longitude*
3. *33° south latitude; 151° east longitude*
4. *48° north latitude; 56° west longitude*
5. *0° latitude; 78° west longitude*

5. Distribute **Handout 10** on topographical maps. Allow students time to answer the questions and have them share their answers to determine whether or not more time is required on this particular skill.

Suggested Responses:

1. *B2*
2. *Radio tower*
3. *Trail*
4. *10 feet*
5. *20 feet*
6. *C6, C7, D6, D7*
7. *Orchard*
8. *Unimproved road*
9. *40 feet*
10. *Minus 10 feet*
11. *50 feet*
12. *C3*
13. *10 feet*

6. For a follow-up activity, place students in small groups and have them brainstorm responses to the following question: What are some ways and/or situations in which you might use some of these skills in your own life or on a job? Remind students that in brainstorming, there is no one right answer—the idea is simply to list as many as is possible.

Suggested Responses:

Reading the text

Grid coordinates with computers

Reading road maps or city maps

Understanding news programs that use maps

Making your own maps

Military maps

Reading or creating blueprints

Enrichment/Extension

1. Locate maps of your state or city and find distances or grid coordinates and map symbols for capitals, state or national parks, state monuments, what roads to take from one city to another, etc.
2. Using the state and city maps, write a set of directions, but do not list the city of departure or arrival. Exchange your questions with another student to see how proficient you each are in locating information on a map.

Name________________
Date________________

Reading Map Grids

Some maps, especially state maps, use map grids for locating places quickly. The grid is made up of numbers and letters to define the squares on the grid. Study the map below, then answer the questions which follow.

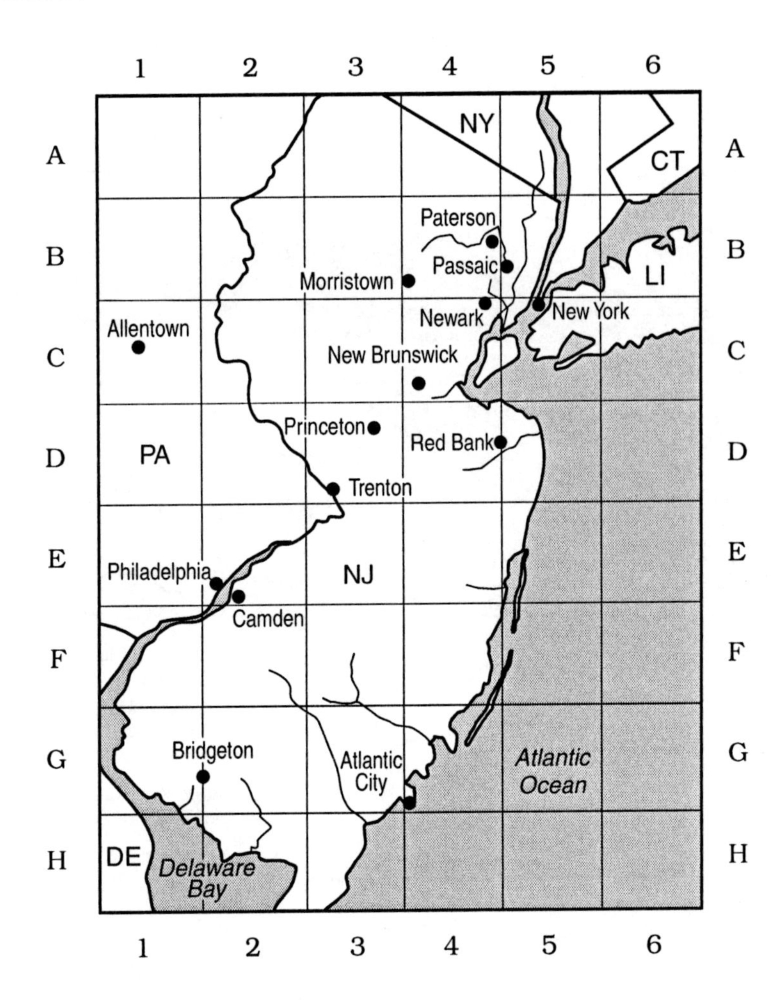

Fig. 3.1.

Questions

1. What cities are located in B4?
2. What are the grid coordinates for Connecticut (CT)?
3. Delaware Bay is located primarily in what two grids?
4. What are the grid coordinates for Philadelphia?
5. New York City sits on the line between which two grid coordinates?

Fig. 3.1. Helen H. Carey, *How to Use Maps and Globes*, (New York: Franklin Watts, 1983), 38, Fig. 18.

Name________________
Date________________

Reading Map Scales

This exercise involves determining the distance on a map by using a map scale. The map below uses a scale when one inch equals approximately 90 miles or 144.8 kilometers. Use a ruler to measure the *direct* distance between features on a map, not as it would show on a road map with turns, and determine how the scale shows distance. For example, if the distance from city A to City B is two-and-one-half inches, then the distance is 225 miles (one inch of 90 miles + half inch of 45 miles = 225 miles). Using the map below, answer the questions that follow.

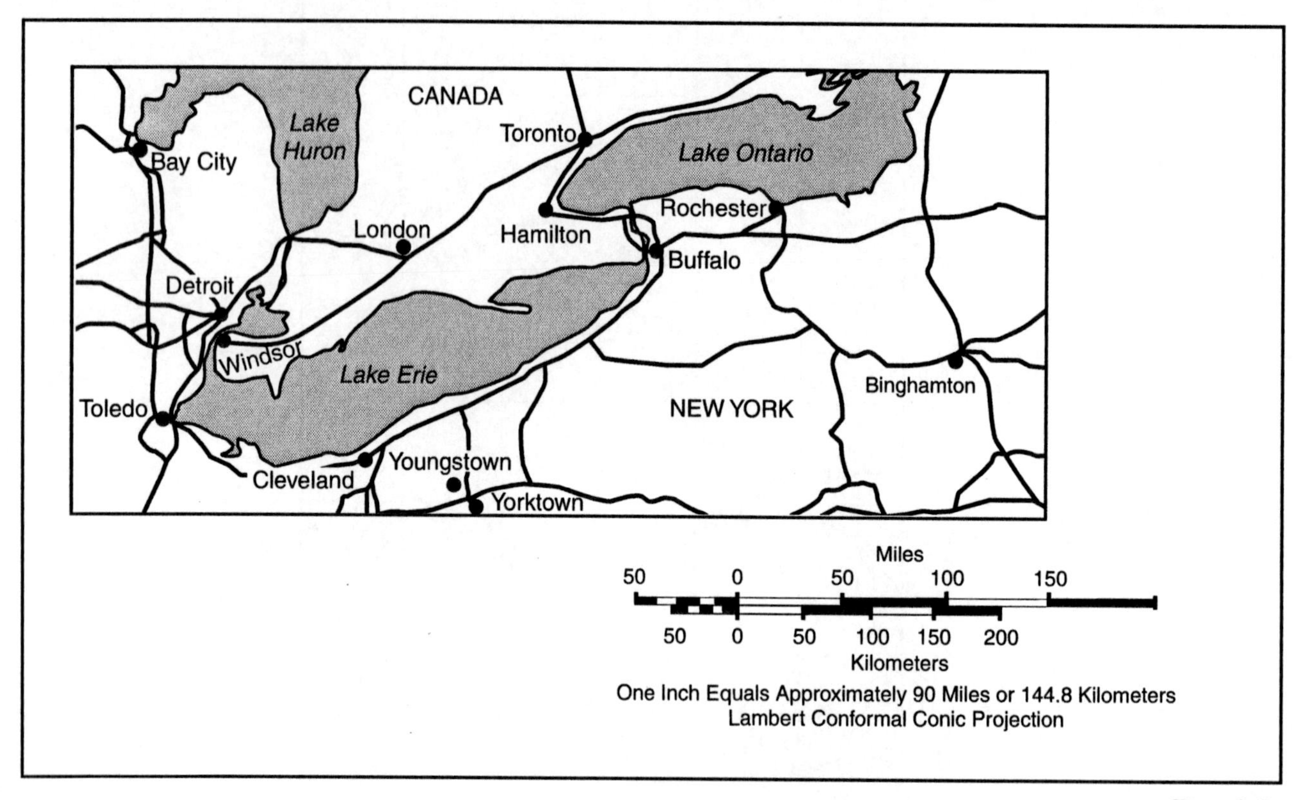

Fig. 3.2.

Questions

1. What is the distance from Cleveland to Toronto?
2. What is the distance from Bay City to Rochester?
3. What is the distance from Youngstown to Buffalo?
4. What is the distance from Toledo to Detroit?
5. What is the distance from Windsor to Toronto?

Fig. 3.2. Adapted from Askov and Kamm, *Study Skills in the Content Areas*, (Newton, Mass.: Allyn and Bacon, 1982), 148, Fig. G-20.

Name________________
Date________________

Longitude and Latitude Exercises

Latitude and longitude are measurements that help us locate places on a map or globe. *Latitude* is distance north or south of the equator. We measure distances in degrees from that reference point. The lines that measure latitude are parallel to the equator so we refer to them as parallels of latitude. *Longitude* is distance east or west of the prime meridian, an imaginary line going through Greenwich, a suburb of London. The lines measuring longitude are called meridians of longitude. Use the accompanying map to answer the questions that follow.

Part A.

Find the places indicated. Unless you are directed to identify a specific country, it is sufficient here to identify the places indicated by either continent or ocean.

1. Locate 10° south latitude and 40° west longitude. Where are you?
2. Locate 10° north latitude and 20° west longitude. Where are you?
3. Locate 10° north latitude and 20° east longitude. Where are you?
4. Locate 30° south latitude and 80° west longitude. Where are you?
5. Locate 30° south latitude and 80° east longitude. Where are you?
6. Locate 60° north latitude and 40° east longitude. Where are you?
7. Locate 60° north latitude and 40° west longitude. Where are you?
8. Locate 20° south latitude and 140° east longitude. Where are you?
9. Locate 40° north latitude and 100° east longitude. Where are you?
10. What *country* would you be in at 60° north latitude and 100° west longitude?
11. What continent would you be on at 40° north latitude and 0° longitude?
12. Where would you be at 40° north latitude and 160° west longitude?
13. Where would you be at 40° south latitude and 160° west longitude?
14. Why is it important to indicate both direction as well as degrees in identifying latitude and longitude?

Part B.

Give the latitude and longitude of each of the following cities.

1. Capetown, South Africa (point A on the map)
   Latitude
   Longitude
2. Buenos Aires, Brazil (point B on the map)
   Latitude
   Longitude
3. Sydney, Australia (point C on the map)
   Latitude
   Longitude
4. St. John's, Newfoundland (point D on the map)
   Latitude
   Longitude
5. Quito, Ecuador (point E on the map)
   Latitude
   Longitude

World Cultures and Geography
Lesson 3
Handout 9 (page 2)

Name________________
Date________________

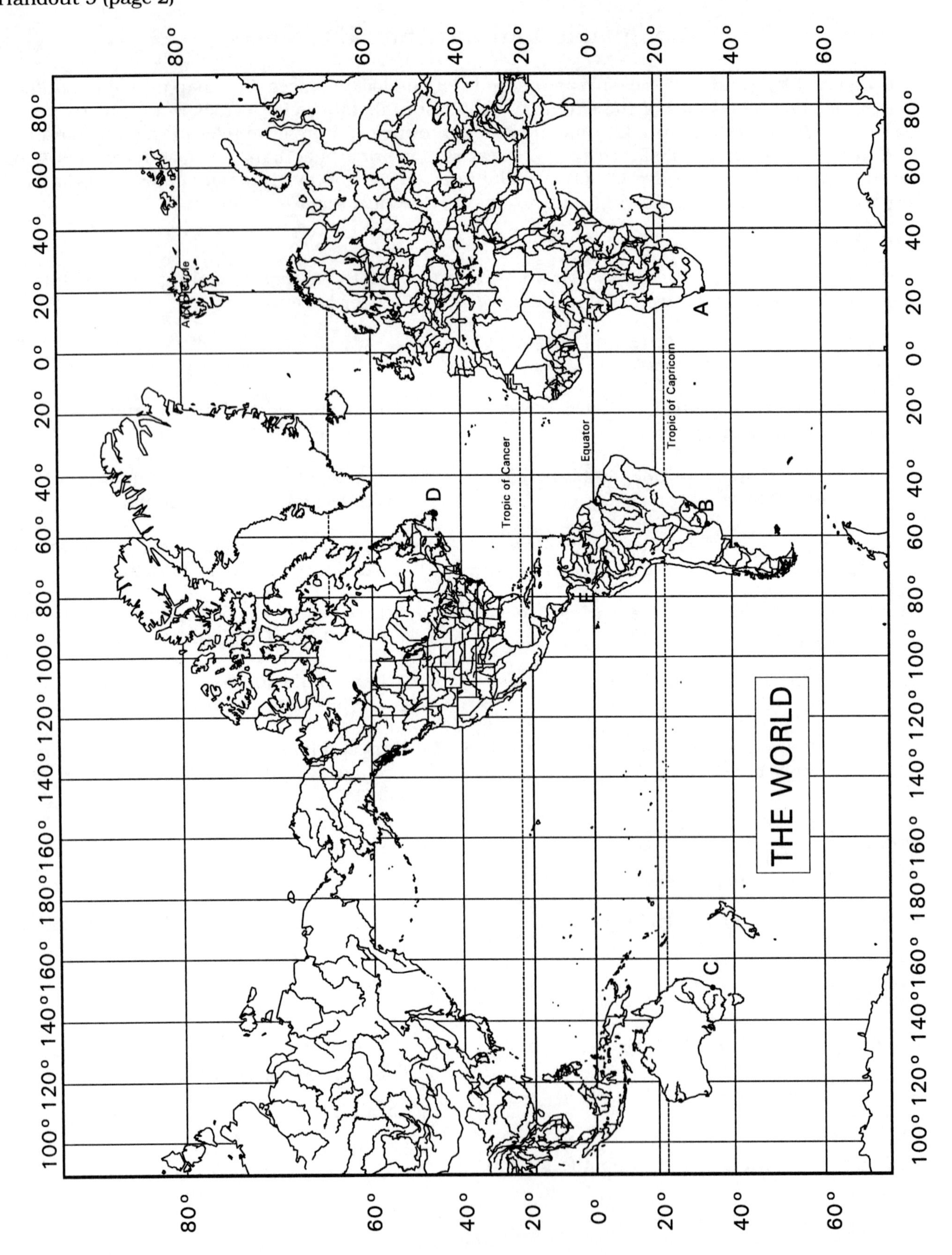

Name____________________
Date_____________________

Reading a Topographical Map

A topographical map shows how the land areas rise and fall. In that way, it combines what a map does (shows features) with what a globe does (shows contours). Like a road map, a topographical map uses symbols to show what features are represented. Study the symbols and the topographical map and review what you have learned about map grids. With this information, answer the questions.

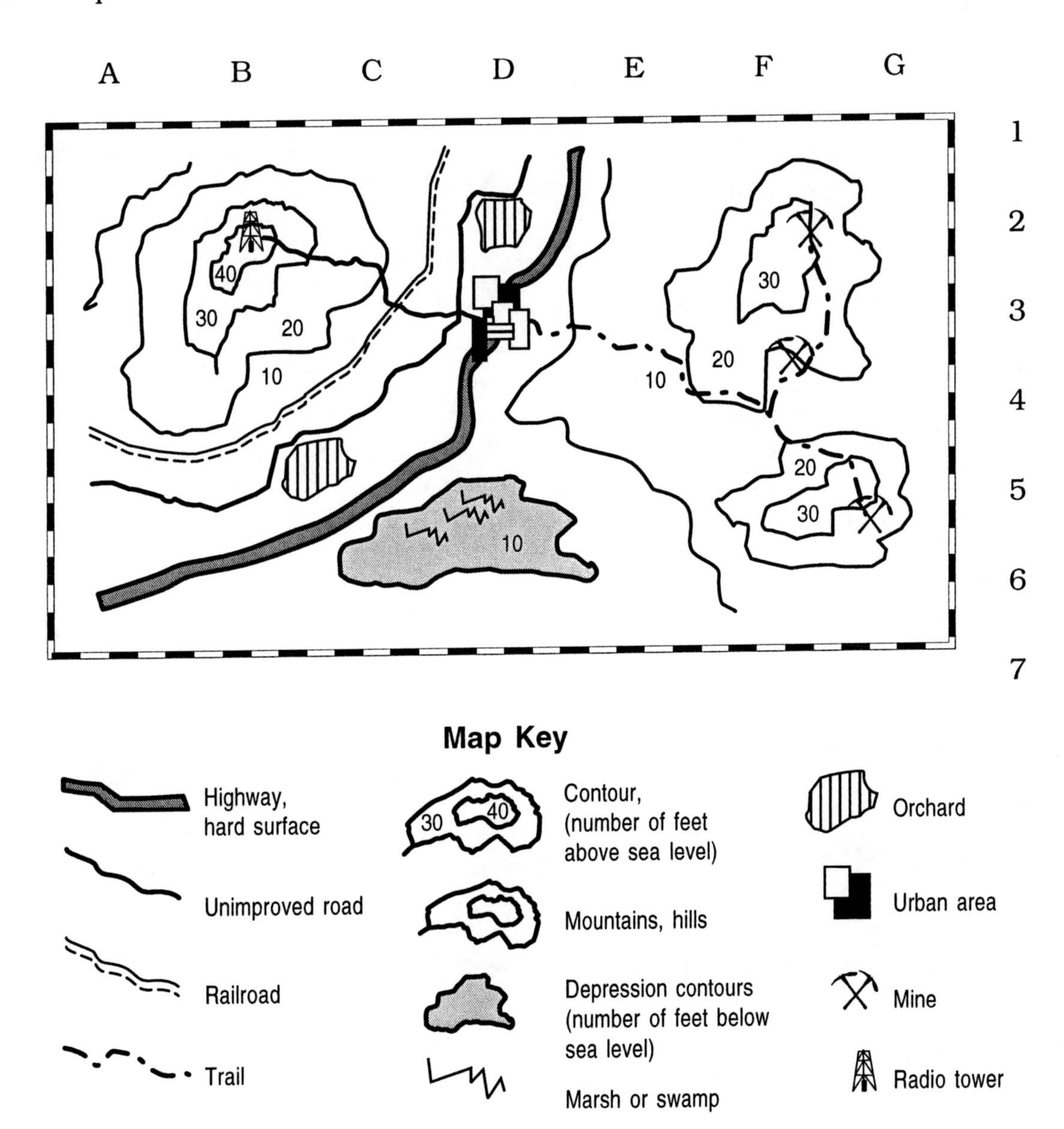

Name________________

Date________________

1. List the grid coordinates for the radio tower.

2. What is located at the highest point on the map?

3. How would you travel to get to the mines?

4. At what height is the railroad?

5. At what height is the mine at grid coordinate G6?

6. Give the inclusive grid coordinates for the swamp.

7. What is located at D2?

8. How would you travel to get to the radio tower?

9. What is the highest point on the map?

10. What is the lowest point on the map?

11. What, then, is the range of heights on the map?

12. At what grid coordinate does the unimproved road cross the railroad?

13. At what height is the trail at grid coordinate E4?

Lesson 4
Using Graphs and Charts to Convey Information

Objectives

- To use graphs and charts to gather information quickly
- To construct graphs from information given in chart form
- To make geographic generalizations based on information in charts and atlases

Notes to the Teacher

Students need to know that charts and graphs can deliver information quickly and can help them understand concepts and relationships. It is important that students know how to use these tools appropriately. Help students understand that this is a shortcut approach to handling many important facts and that newspapers, television, magazines, as well as other businesses, utilize charts and graphs frequently when presenting materials to their audiences. Like a picture, these graphic aids allow for quick delivery and understanding. Reinforce these skills when using textbooks, newspapers, etc., by reading charts and graphs as they are encountered in these content areas.

Students read a chart showing temperatures for a given day and answer questions. They also read a chart giving the annual average temperatures and precipitation for twenty cities scattered around the world. They use an atlas to locate the cities, and along with the chart, use the information to make some geographic generalizations. Students create a bar graph using the chart showing the height of ten mountains around the world. A third activity has students creating a circle graph using census information from Hawaii. Students may need to review how percentages are calculated.

Procedure

1. Show students samples of graphs and charts. Be sure that line, bar, and circle graphs are represented. Ask students how they have used graphs and charts to get information (most will have used graphs and charts). Have students define the various types.

   Suggested Responses:

   a. Bar graph—*shows changes by using thick lines or bars which may be black, white, gray, or color-coded*

   b. Circle or "pie" graph—*shows relation of parts to the whole*

   c. Line graph—*uses narrow lines to show numerical or directional changes*

   d. Chart—*tables of numbers present information at a glance*

2. Have students read the chart and answer the questions on **Handout 11.** Review students' responses and be sure they understand how to read the chart.

   Suggested Responses:

   1. *Acapulco*
   2. *Cairo*
   3. *Sydney*
   4. *Stockholm*
   5. *London 67/56; Copenhagen 76/53; Geneva 87/60; Rome 80/61*
   6. *London, cloudy; Copenhagen, sunny; Geneva, sunny; Rome, sunny*

3. Distribute **Handout 12** and explain that this chart has the yearly averages found in these cities from around the world. The averages were obtained based on thirty years of record keeping at the various weather stations in these cities. Supply each student with a world atlas and ask students to work in groups of three to answer the questions and to form geographic generalizations based upon the information found in the chart. Review correct responses. Share and discuss the generalizations.

Suggested Responses:

1. *La Paz, Bolivia 12,001 ft.*
2. a. *Sydney, Australia 114°*

   b. *South*

   c. *January*
3. Northern cities—*Algiers, Addis Abba, Amsterdam, Bogota, Bombay, Caracas, Dublin, Geneva, Havana, Jerusalem, Lagos, La Paz, London, Moscow, Paris, Tokyo, Toronto*

   Southern cities—*Auckland, Capetown, Sydney*

   North of the equator—*January is winter and July is summer*

   South of the equator—*January is summer and July is winter*
5. *As the elevation increases, the temperature decreases.*
6. *It is warmest at the equator at sea level. The farther from the equator one travels either north or south latitude, the cooler the temperature unless modified by mountains, water bodies, or elevation.*

Sample Geographic Generalizations:

At high elevations, temperatures are moderate even near the equator.

Cities located in the interior of middle latitudes have greater temperature extremes.

Cities located in coastal regions report cooler, more moderate temperatures.

4. Have students use **Handout 13** to develop a bar graph of mountain heights using the information found on the chart. Help students with the directions and emphasize that they are to order them on the graph from smallest to highest. Before starting the graph, have students number the mountains in the proper order along the edge of the chart to avoid mistakes. Instruct students to start the vertical column of the graph at 5,000 feet and go up in intervals of 5,000 feet ending with 30,000 feet. Be sure students place the mountain names on the horizontal bar. Suggest that they add color to their graphs. Display completed graphs.

5. Distribute **Handout 14** and have students make circle graphs with the information in the chart about Hawaii's population. Remind students that they must figure the percentages of population for each of Hawaii's counties, and write the percentages next to the population figures on their charts. (Note: Be sure students understand that the percentage is found by dividing the part by the whole: 120,317 divided by 1,108,099.) Round off the decimals to the nearest hundreds place.

   Suggested Responses:

   Hawaii .1085796 = 11%

   Honolulu .7546536 = 75%

   Kauai .0461845 = 05%

   Maui .0905821 = 09%

   When students have the percentages correctly calculated, have them divide their circle graphs accordingly. Instruct students to mark each section with the county name, population, and percentage figures. Suggest that they add color to their graphs for emphasis.

6. Display students completed graphs. Continue throughout the year to review the use of charts and graphs whenever they appear in textbook or supplementary instructional material.

Enrichment/Extension

1. Look for examples of different kinds of graphs and create a classroom display. Be sure you can explain/interpret the graphs you contribute to the display.
2. Create your own graphs to illustrate the statistics for your team sports at school or for professional sports teams; weather data; your grades; etc.

Name________________
Date________________

What Charts Can Tell You

Complete the following.

1. What city had the highest temperature on June 24?
2. What city had the third highest?
3. What city had the lowest temperature?
4. What city had the fifth lowest?
5. What temperature changes would you experience if you flew from London to Copenhagen to Geneva to Rome?
6. What kind of weather would you have had in each of these cities?
7. Locate each of these cities on the wall map and on the map of Europe in your textbook.

World Temperatures June 24 (One day)

City	Temp.	City	Temp.
Acapulco	92/78 pc	London	67/56 c
Amsterdam	78/58 s	Madrid	80/59 s
Athens	84/68 s	Manila	80/74 sh
Auckland	57/43 s	Mexico City	74/56 sh
Barbados	89/75 pc	Moscow	58/49 r
Beirut	78/68 s	Nassau	85/74 t
Berlin	71/53 s	Oslo	74/64 s
Bermuda	78/70 c	Paris	78/62 pc
Brussels	80/60 pc	Peking	82/62 pc
Buenos Aires	58/50 pc	Rio de Janeiro	80/68 s
Cairo	88/70 s	Rome	80/61 s
Copenhagen	76/53 s	San Juan	89/77 pc
Dublin	61/51 sh	Seoul	82/71 c
Edmonton	73/61 pc	Stockholm	65/46 s
Freeport	86/74 t	Sydney	59/42 s
Geneva	87/60 s	Tokyo	76/69 sh
Helsinki	62/45 s	Vancouver	67/57 c
Hong Kong	86/83 sh	Vienna	72/57 c
Jerusalem	79/60 s	Warsaw	64/45 pc
Lima	65/56 pc	Winnipeg	59/50 pc
Lisbon	70/64 pc		

Numbers are high/low temperatures; letters indicate:

s: sunny	c: cloudy	pc: partly cloudy
sh: showers	r: rain	t: thunderstorms
i: ice	sn: snow	sf: snow flurries

Name__________________
Date__________________

Weather Information Unit

Use this chart and an atlas to answer the questions that follow.

Annual Average Temperature/Precipitation

Weather station	Elevation feet	January maximum	January minimum	July maximum	July minimum	Extreme maximum	Extreme minimum	Average precipitation
Addis Abba, Ethiopia	8,038	75	43	69	50	94	32	48.7
Algiers, Algeria	194	59	49	83	70	107	32	30.0
Amsterdam, Netherlands	5	40	34	69	59	95	3	25.6
Auckland, New Zealand	23	73	60	56	46	90	33	49.1
Bogota, Columbia	8,355	67	48	64	50	75	30	41.8
Bombay, India	27	88	62	88	75	110	46	71.2
Capetown, South Africa	56	78	60	63	45	103	28	20.0
Caracas, Venezuela	3,418	75	56	78	61	91	45	32.9
Dublin, Ireland	155	47	35	67	51	86	8	29.7
Geneva, Switzerland	1,329	39	29	77	58	101	-1	33.9
Havana, Cuba	80	79	65	89	75	104	43	48.2
Jerusalem, Israel	2,654	55	41	87	63	107	26	19.7
Lagos, Nigeria	10	88	74	83	74	104	60	72.3
La Paz, Bolivia	12,001	63	43	62	33	80	26	22.6
London, England	149	44	35	73	55	99	9	22.9
Moscow, Russia	505	21	9	76	55	96	-27	24.8
Paris, France	164	42	32	76	55	105	1	22.3
Sydney, Australia	62	78	65	60	46	114	35	46.5
Tokyo, Japan	19	47	29	83	70	101	17	61.6
Toronto, Canada	379	30	16	79	59	105	-26	32.2

Name__________________
Date___________________

1. Which city is located at the highest elevation?

2. a. Which city has reported the highest temperature?

   b. Is it north or south of the equator?

   c. What month do they usually experience high temperatures?

3. List all the cities north of the equator? List all the cities south of the equator. Tell whether the January temperatures are summer or winter measurements.

4. During your holiday break in December, to which city would you like to go to spend your vacation? Why?

5. Explain how elevation affects temperatures. Give examples from the chart.

6. How does the distance from the equator (degrees of latitude) affect temperatures? (Use an atlas to determine at which latitude these cities are located.)

Name________________
Date________________

Making a Bar Graph

Draw a graph showing the various heights of the mountains listed in this chart. Arrange them in order by height—smallest to largest.

Mountains		**Height in Feet**
Mount Everest	Nepal	29,208
Mount McKinley	Alaska	20,320
Popocatepetl	Mexico	17,887
Mount Ranier	Washington	14,410
Aconcagua	Argentina	22,834
Mount Kilimanjaro	Tanzania	19,340
Mont Blanc	France	15,771
Vinson Massif	Antarctica	16,864
Mount Hood	Oregon	11,235
Kosciusko	Australia	7,310

Name________________
Date________________

Making a Circle Graph

Use the information in the chart to create a circle graph to show the percentages of Hawaii's population in each county. First, figure what each county's population percentage is. Then divide the circle into sections using those figures. Mark each section with its county name and population figures.

Hawaii's Counties	**Population**	**Percentage**
Hawaii	120,317	
Honolulu	836,231	
Kauai	51,177	
Maui	100,374	
Total population	1,108,099	

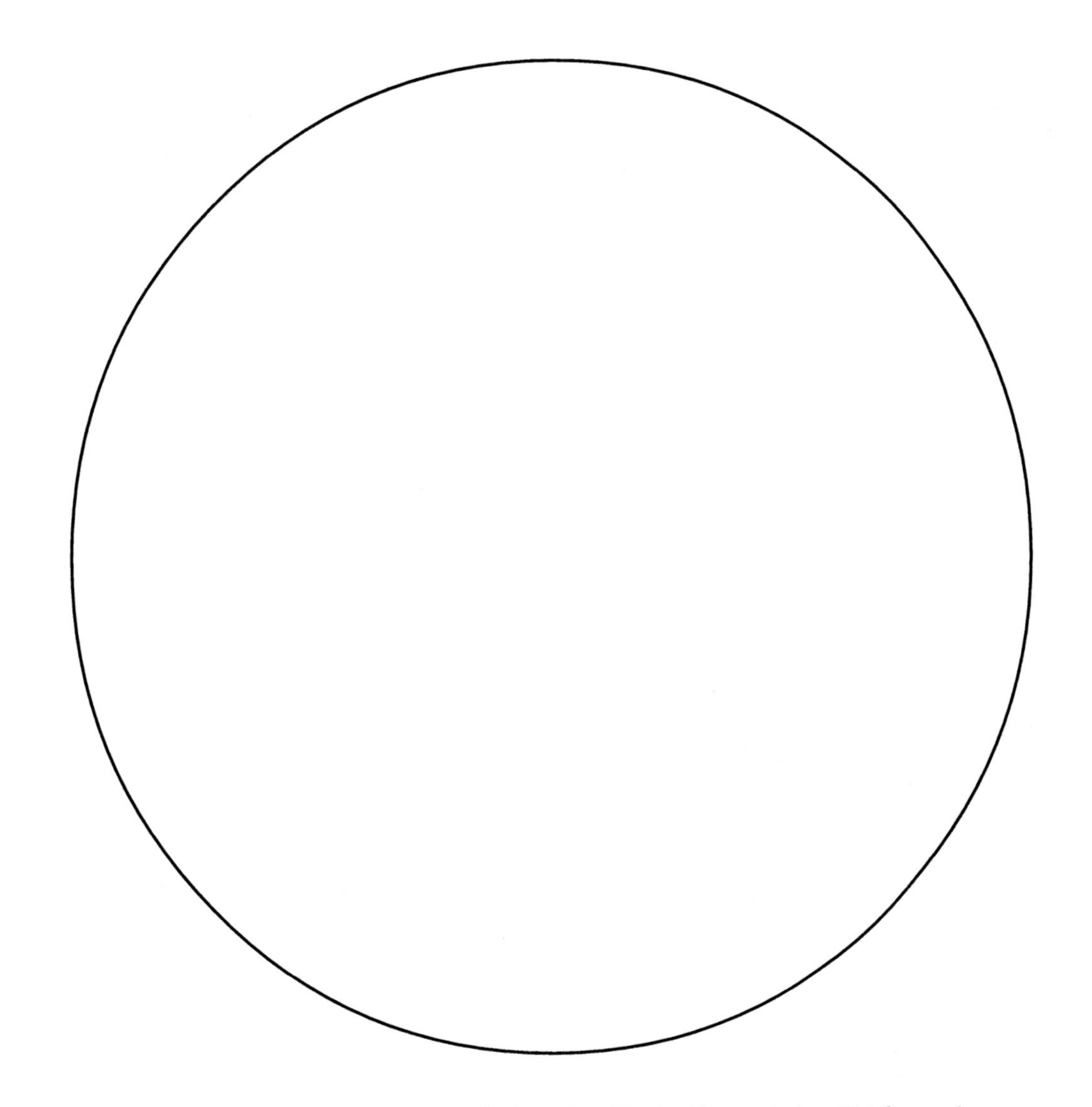

Lesson 5
What Is Culture?

Objectives

- To describe culture in geographic terms and context
- To arrive at a working definition of culture

Notes to the Teacher

In its broadest terms culture involves all the institutions, beliefs, arts, and behavior patterns characteristic of a community or population. Modern geographers generally agree that humanity has shaped its environment rather than that environment determined human development. From earliest times humans have learned to improve their surroundings and harness natural resources to their service. Unfortunately, humans have also abused the earth and squandered its seemingly boundless supply of earth, air, water, and natural resources. The generations who spawned the industrial revolution, and especially our present generation, have been the greatest spendthrifts of time and resources.

As we examine the world through the geographer's eye, we ask why people settled where they did and how and why they defined their living space. Their pattern of living showed consistent characteristics which identified them as family, clan, tribe, or state. Within that framework, people developed their way of life on a systematic base of prevailing values and beliefs; language; social, economic, and political norms; and social control through law and traditional values and practices which each generation passed on to the next.

Students develop a working definition of culture and study its application in their own country, in an imaginary setting, and in ancient Greece. Students also develop hypotheses regarding patterns of cultural development and the universality of culture, students can test these hypotheses in their later studies in the course.

Procedure

1. Have ten students use dictionaries to look up the following terms and write definitions on the chalkboard.

1. culture	6. language
2. society	7. education
3. economy	8. government
4. politics	9. fine arts
5. philosophy	10. environment

2. Help students analyze and discuss the dictionary definitions. Have students check the glossary in their text for extension or limitations of the terms.

3. Ask students how the terms are related. Develop with them a visual (*e.g.*, web) to illustrate connections. Students should recognize that a culture develops within a particular environment and that society, economy, politics, philosophy, language, education, government, and fine arts are aspects of a given culture.

4. Ask students to brainstorm as many components as possible of a culture which they know. List them on the board.

   Suggested Responses:

   a. *Music, art, literature*

   b. *Prevailing modes of dress*

   c. *Routine living habits*

   d. *Food preferences*

   e. *Architecture of houses, public buildings*

   f. *Layout of fields and farms*

   g. *Systems of education, government, law*

5. Have students use the visual developed in procedure 3 to write their own definition of "culture." Have them note similarities in their individual definitions and try to arrive at a class definition.

6. Pair students briefly to develop a quick characterization of their own culture in terms of their definition. Discuss similarities and differences students noted and possible reasons for differences. Share with the class information in the Notes to the Teacher.

7. Distribute **Handout 15**. Have students read the story silently and answer the questions. Then, in their small groups, have students deduce how the culture described developed in that way. Compare group responses.

   Suggested Responses:

   1. *Community, water, food, shelter, clothing*
   2. *Language, law, social customs, fire*
   5. *They reflected their own "roundness" in their way of life.*

8. Distribute **Handout 16**. Read the selection aloud and discuss responses to the questions.

   Suggested Responses:

   1. *Boats, farming, fishing, harnessing beasts of burden, building houses*
   2. *Language, developed thinking, mastery over some diseases*
   3. *Death*
   4. *Sense of optimism that he can control his destiny*
   5. *Believes gods side with righteousness*
   6. *May have begun destruction of earth through farming techniques*
   7. *Banished from community*

9. To conclude the lesson, discuss possible generalizations or hypotheses about culture and patterns of cultural development that students might make and test in later lessons. For instance, students might note that from the earliest times, communities all developed a need for the same aspects of culture.

Enrichment/Extension

1. Join with students from different ethnic groups to report on customs of inherited cultures, including national costumes, food, music, art, architectural styles. Interview older members of the ethnic groups, such as grandparents, for authentic information.
2. If you have visited other countries share your observations and experiences.
3. Do a library report on the culture of one of the following groups. Be prepared to share your findings as an oral report.

   Stone Age People (2½ million years ago)

   Neanderthal People (70,000 B.C.)

   Cave People (where art began) (15,000 B.C.)

   The First Farmers (10,000–2,500 B.C.)

   The Metal Workers (6,000 B.C.)

   An American Indian Tribe (Aztec, Mayan, Inca)

   Recommended Sourcebooks

   Aztecs: Reign of Blood and Splendor, Lost Civilization Series. Alexandria, Va.: Time-Life Books, 1992.

   The Human Dawn, Time Frame Series, Time-Life Books, Alexandria, Va.: Time-Life Books, 1992.

   Incas: Lords of Gold and Glory, Lost Civilization Series. Alexandria, Va.: Time-Life Books, 1992.

   The Last Two Million Years, London: Reader's Digest Association Limited, 1973.

   The Magnificent Maya, Lost Civilization Series. Alexandria, Va.: Time-Life Books, 1993.

Name________________
Date________________

A Round World

Read the story silently. Answer the questions in your notebook. Be prepared to share your responses with your group.

Once upon a time, long ago and far away, there was a large island named Obo which was perfectly round and set in the middle of a fresh water lake. The only trees were coconut palms which grew in abundance. There was one mountain which dominated the northern section of the island from whose round top on a clear day one could see the mainland.

The people, known as Obotes, were very round also, like the Pillsbury flour man in the television commercial. They built round houses with round roofs, round windows, and even round doors. Their paths were laid out in concentric circles whose center was their round house. The larger paths around the island were also circular. They ran through the coconut grove and around the base of the round mountain. In the center of their village, they had a round marketplace where they could exchange coconuts for fresh water fish which had been caught by their fishermen from round boats fashioned from many coconut shells, lined with palm leaves.

The round coconuts were the principal resources of these islanders. Workers picked them, women drained the milk into empty shells, used the meat of the nut to feed their families, and wove the coarse fibers into garments. The men used the strong outer shells for building shelters and boats.

Since the coconut was the staple of their lives, they worshipped the spirit of the trees which produced it. The high priest of the tribe, the Obobo, selected the best coconuts for drums to be used in their festivals of worship. He also selected the roundest young men to be the drummers and dancers. The young women of the tribe made the elaborate, round masks for the dancers. They also carved designs on the doors of their houses and on containers for food.

When an Obote died, the tribe built a round mound over his remains and sang a dirge of musical rounds at his grave.

Altogether, we might say, the Obotes lived a well-rounded life!

1. What components of a culture do the Obotes have?

2. What elements of a culture are not mentioned in the story but are implied?

Name__________________
Date__________________

3. Discuss the kinds of law that they might have. Would their laws stress equality? Why or why not?

4. What would their language be like? Would it be full of open vowel sounds like a, o, u, and combinations of these? Make up some words that the Obotes might use for their needs of daily life.

5. Did the Obotes control their environment or did the environment control them? Give reasons for your answer.

6. Do you think that someday an Obote would come along who would notice the tall, slender, perpendicular trunk of the palm trees and introduce new ways of building to his fellow-islanders? How would these ideas be received? Why?

Name________________
Date________________

What the Greeks Believed

Read the following and be prepared to discuss the questions that follow.

Chorus

Many the wonders but nothing walks stranger than man.
This thing crosses the sea in the winter's storm,
making his path through the roaring waves,
And she, the greatest of gods, the earth—
ageless she is, and unwearied—he wears her away
as the ploughs go up and down from year to year
and his mules turn up the soil.

Gay nations of birds he snares and leads,
wild beast tribes and the salty brood of the sea,
with twisted mesh of his nets, this clever man.
He controls with craft the beasts of the open air,
walkers on hills. The horse with his shaggy mane
be holds and harnesses, yoked about the neck
and the strong bull of the mountain.

Language, and thought like the wind
and the feelings that make the town,
he has taught himself, and shelter against the cold,
refuge from rain. He can always help himself.
He faces no future helpless. There's only death
that he cannot find an escape from. He has contrived
refuge from illnesses once beyond all cure.

Clever beyond all dreams
the inventive craft that he has
which may drive him one time or another to well or ill.
When he honors the laws of the land and the gods' sworn right
high indeed is his city; but stateless the man
who dares to dwell with dishonor. Not by my fire,
never to share my thoughts, who does these things.[1]

This poem was written in the fifth century B.C. by Sophocles as a song between the scenes of his tragic drama, *Antigone*. He expands the thought of an early Greek thinker, Pythagoras, who said: "Man is the measure of all things."

[1]David Grene and Richard Lattimore, eds. *The Complete Greek Tragedies: Sophocles Vol. 2.* (Chicago: University of Chicago Press, 1959), *Antigone*, lines 345–372.

Name________________
Date________________

Discuss the following questions in your group.

1. Find at least five ways the poet shows that humans have gained mastery over their world.

2. What great things beyond the physical activities described in the poem has humanity taught itself?

3. What is the one element of life humanity has not yet conquered, then or now?

4. What general attitude of the author toward the environment can you detect?

5. How does honoring the gods affect his well-being?

6. Is there any hint that humanity's use of its world may not always be good? What is it?

7. What is the speaker in the poem saying of those who do ill?

Lesson 6
Ethnocentricity

Objectives

- To examine the concept of ethnocentricity
- To survey the stereotypes and biases of ethnocentricity

Notes to the Teacher

In order for students to appreciate the cultural diversity that is part of a study of world cultures, it is helpful to assist them in emptying their minds of the cultural perspective that they bring to such a study. Since all people are a product of their cultures, each person has an enthnocentric view of the world (ethno = cultural group; centric = at the center). In and of itself, ethnocentricity leads to loyalty to, and pride in, one's own nation and culture; yet if one looks at the world only through these glasses, one never learns to appreciate the diversity of the family of humankind.

In order to prepare students for this lesson, it helps to establish common understandings of terms to avoid debating terms when the focus of the lesson should be on concepts.

ethnocentric—reference to the idea that one's own culture is better or more correct than another's culture

nationalism—a value judgment that one's own nation is not only better, but that its values and traits should be promoted over all others

stereotypes—over-simplified, commonly-held opinions of a person or group; often a composite of traits

bias—personal, distorted judgment that influences objective perception

prejudice—unreasonable attitude or bias directed against a group or culture based upon supposed characteristics

discrimination—an action based on a prejudice

Procedure

1. Introduce the lesson using material in the Notes to the Teacher. Focus on definitions of key terms. This can be done in several ways—by having selected students look up definitions in a dictionary or glossary, by having students share how they perceive the terms and summarize the group definitions on the board, or by giving definitions to students from an overhead projector or chalkboard.

2. Distribute **Handout 17**. This handout makes students aware that they have preconceptions of which they are unaware, and that these preconceptions color their perception. After allowing each student a few minutes to record character traits and qualities that they associate with the names, give them the following list of traits associated with names as collected from surveys.

   Amy = active

   Jennifer = youthful, yet old-fashioned

   Lisa = well-liked

   Elena = bright, quiet

   Nicole = average

   Carlos = strong, manly

   Michael = very popular, extremely manly

   John = trustworthy, passive yet manly

   Christopher = intelligent, hard-working

   Jason = hugely popular

   Have students share their answers to the questions at the end of the handout. It is important to have students understand that there are no "right" answers here, but only results of surveys. Consider "surveying" the class to see if they have a group perception of the traits that are dominant. Answers to the questions at the end of the handout will no doubt vary, yet it may be a worthwhile educational experience to have students share their responses to the questions with the class.

3. Distribute **Handout 18**. Allow students time to complete the matching.

   Suggested Responses:

1. g	*6. j*
2. d	*7. b*
3. i	*8. h*
4. a	*9. f*
5. c	*10. e*

4. Have students then complete the questions at the end of the handout. Review their responses and use them as a springboard for discussion.

   Suggested Responses:

   3. Students should arrive at a similar conclusion—that they are now aware that they carry with them preconceptions (stereotypes, biases, prejudices) that impede a free and open influx of information. Hopefully, students have also come to realize that this is a result of their ethnocentricity. Students' own experiences may also have prompted them to conclude that there are important exceptions to commonly perceived stereotypes. Most importantly, they should see that awareness of their attitudes should help them avoid discrimination or action based on inappropriate attitudes.

5. To conclude the lesson, model the process of effective writing by creating with the class a good paragraph on the topic of ethnocentricity. Incorporate the other key terms of the lesson: nationalism, stereotypes, bias, prejudices, and discrimination. Have students write the paragraph in their notebooks as you and they develop it together.

Enrichment/Extension

1. Construct a survey on stereotypes within your own school community, or the community at large.
2. Compile a list of job or social role stereotypes from television programs or movies.
3. Bring to class advertisements from magazines and share what you observe as stereotypes used in advertising—ages or sexes or ethnic groups in relationship to the products being sold.

World Cultures and Geography
Lesson 6
Handout 17

Name__________________
Date__________________

What Is in a Name?

Look at the list of ten names in the chart. In the column marked Qualities and traits, write down the kind of personality traits that you associate with people who have those names (example: Tammy—pert, mischievous). When you have finished, your teacher will tell you what qualities and traits have been identified in surveys. After you have written down the name's qualities from the survey, answer the questions at the bottom of the page. Be prepared for class discussion.[1]

Name	**Qualities and traits** (how you see them)	**Traits as surveyed** (how others see them)
Amy		
Jennifer		
Lisa		
Elena		
Nicole		
Carlos		
Michael		
John		
Christopher		
Jason		

Questions

1. On how many traits that you listed did you agree with the survey? Disagree?

2. Did you find that for one or more of the names you know of actual people who do *not* match the survey's responses? Which ones? Why?

[1]Adapted from Christopher P. Anderson, *The Name Game*, 1977. Published by Peoples Weekly, New York.

Name________________
Date________________

How Do You View?

In the chart below, match the nations/cultures listed with character traits commonly associated with those cultures. When you have finished, your teacher will supply you with the answers typically given. Then, compete the questions which follow the chart.

Nations/cultures	Character traits of the nations/cultures
_____1. Italians	a. Materialistic, social-climbers
_____2. French	b. Industrious, high-tech
_____3. Arabs	c. Boorish, stuffy
_____4. Americans	d. Sophisticated, lovers
_____5. British	e. State-controlled robots
_____6. Libyans	f. Frugal, penny-pinching
_____7. Japanese	g. Quick-tempered, passionate
_____8. Congolese	h. Lazy, primitive
_____9. Scots	i. Tremendously wealthy
_____10. Russians	j. Terrorists

Questions

1. How many of your answers matched those suggested by your teacher? How many did not?

2. Can you identify examples that disprove these stereotypes as inaccurate?

3. What has this chart exercise said to you about your ethnocentric stereotyping and biases?

Lesson 7
Canvassing Your Community

Objectives

- To identify the elements that comprise a community
- To make decisions about a community based on its human-environment elements
- To investigate the history of the community
- To construct a map of the community identifying key elements

Notes to the Teacher

To study their community, students need to develop a working definition of a community. Once they have defined it, students should begin rehearsing the community in which they live—doing a study of its history, geography, culture, available resources, economic ventures, governmental structure, and avenues for citizen participation. Useful resources for students to look to for information are the local chamber of commerce, local weather bureau, library, local people who have lived their whole lives in the community, city administration offices, state highway department (maps), etc.

Once students have developed this information base, it is important for them to relate their community with others in the United States and in other parts of the world with regard to geography, history, economics, and culture.

Students research their community using a survey sheet and answering the questions on the sheet. Students draw a map of their community marking areas that are important to them individually and those that are important collectively. Finally, students read a scenario, about a community named Paradise, that requires them to make important community decisions. Thus, students practice citizenship participation.

Procedure

1. Show students some pictures of your community—local buildings, people, activities, businesses, etc. Ask students to define what they believe a community is. Entertain all their suggestions until you can construct a class definition. Include elements such as common interests, ethnic groups, people working cooperatively to achieve common goals, environment—plants, animals, topography, etc. (*A community is a place where people live and work together.*)
2. Write the consensus definition on the board. Direct the attentions of the class to their own community. What languages are spoken? What is the weather and climate like? What customs are practiced? What is their location? What types of housing exist? What is the population? What ethnic groups are found there? What are the occupations? What types of transportation are available? What economic resources are available?
3. Distribute **Handout 19**, and instruct students to begin to research some basic information about their community. (This handout is just a base, feel free to add other elements.) Ask students to interview at least three older adults, who have lived a long time in the community, about how the community has changed over time—what has changed in their lifetimes and what has stayed the same.
4. Have students use **Handout 20** to draw a map of the community, filling in the details as described on the sheet. Students must make a key to indicate the meaning of their symbols. Remind students that color coding enhances maps.
5. Debrief the class with information that students have gathered. Note the changes that have occurred within the community over time. Ask the class to predict what changes they think may occur in the future—50 years? 100 years?

   Have students describe the effects humans have had on the natural environment. Has the terrain been altered? How? Explain. Have the plants and animals in the natural environment changed over time? How? Explain. Has the make-up of the population changed over time? How? Explain. How does the natural and cultural environments of a community affect that community?

Have students relate issues of importance to their community and compare those issues with other communities within their own region or around the world. Do they share commonalities? What differences do students see?

6. Assign students to one of the five groups listed on **Handout 21**. Once students are in groups, distribute the handout. Have them read the scenario about Paradise and then determine what stand they will take on the issue based upon how they believe their group would react. The issue to consider is whether to build a plant near Clear River—which might possibly pollute the river, a source of rainbow trout, recreational fishing, and tourism—or allow the senior citizen group to put up housing units for seniors in which the group guarantees NO pollution. Have students list the costs and benefits of their decisions within their groups and then present their recommendations to the class, making an effort to persuade the class to their way of thinking. Remind students that a major highway link will benefit the community by bringing in more tourists and will allow the transporting of goods for sale in other communities. It will also disrupt the natural environment, and animal habitats will be affected as well as people who will have to move when their homes are in the path of the highway. Finally, after all recommendations have been discussed, put the issue to a class vote.

Enrichment/Extension

1. Invite some local people who have been recognized for their contributions to the community to speak to your class.
2. Survey different cultures represented within your community by listing different ethnic restaurants within the community. Bring in sample menus and discuss the countries where these cultures originate. Perhaps the class could arrange to visit one or two of these restaurants to sample the foods. Remind students that just because something is different does not mean that it is less valuable than those familiar things.
3. Build a model of your community from boxes and cardboard. All important downtown buildings should be present. Select a community from another country and build a model of it. Research how that community operates. Construct a Venn diagram which compares the two communities and display it along with your models.

Name__________________
Date__________________

Community Canvas

Research your community and answer these questions.

When was the community settled?

Why did they come?

Why did they choose this particular place to settle?

Who came to settle here?

How fast did the community grow?

What ethnic groups are represented in this community?

How did the early settlers in the community help improve the natural and cultural environment?

What famous citizens came from the community? Why are they famous?

Name________________
Date________________

Mapping Your Community

Draw a map of your community to show the following: the main streets, where you live, major stores, rivers, your school, city hall, library, parks, important businesses, and factories. Develop a key showing what symbols you have used for these places.

Name________________
Date________________

Living in a Community Means Making Choices

A community is a group of people with common interests who share a common environment. The environment is the natural and human-made things around them. The way different groups live is their culture. Culture includes the type of food, housing, clothing, religion, language, government, etc., that a group uses. In the story below, you must make some decision about a community. You will be assigned to work in one of the groups listed. Pretend that this is your community and make good choices that will benefit your community.

Groups: Business Group

Environmental Protection Group

Chamber of Commerce

Senior Citizen Group

City Council

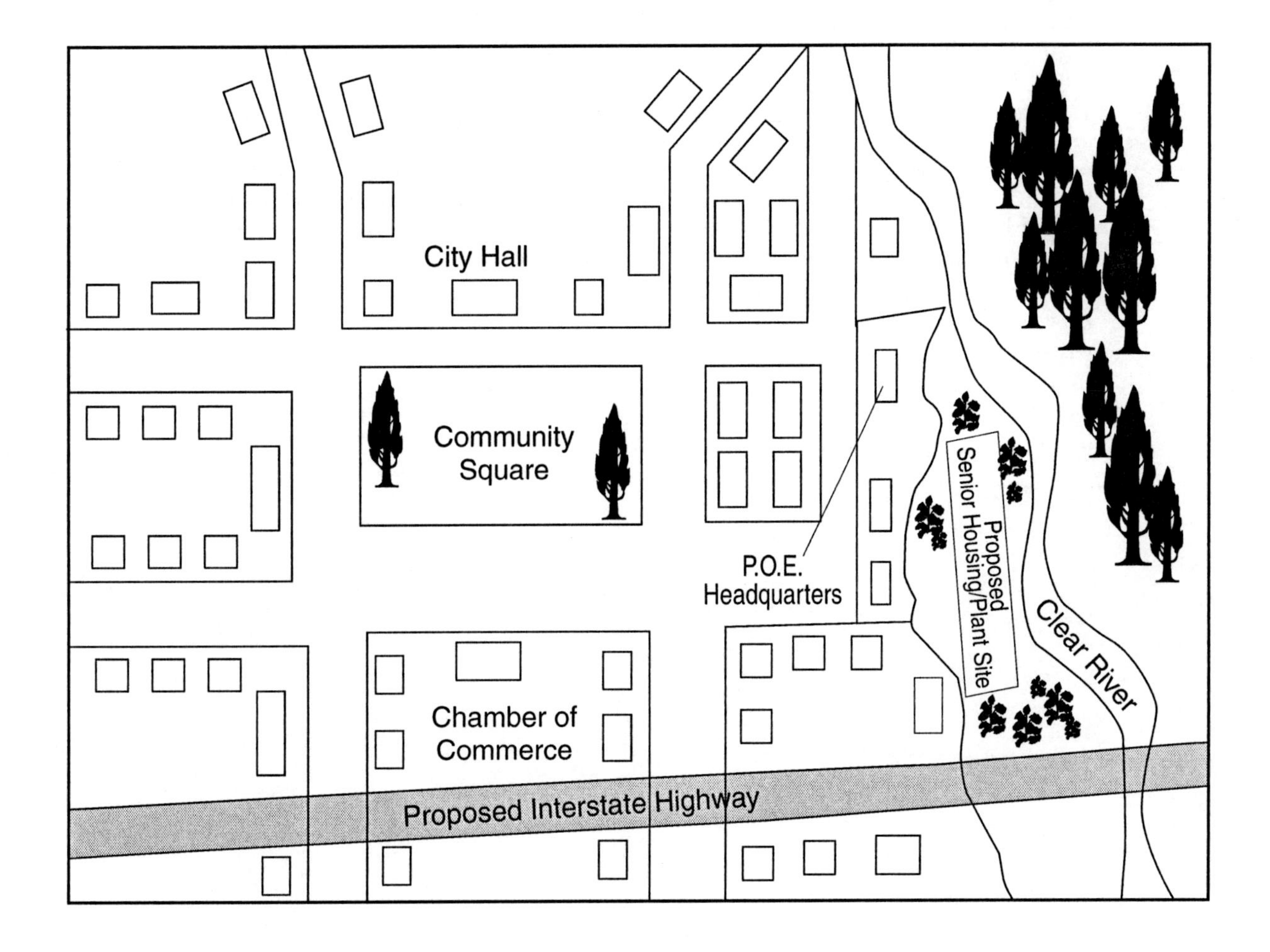

Name________________
Date________________

Paradise—A Story about a Community

Paradise is a small community with a mayor and a city council who want to see it grow in size. A business firm has offered to come in and build a new plant in their community. It will mean many new jobs for the people who live there, but the same firm has a plant 200 miles away and the people there say it pollutes the air and water in their town.

The firm wants to locate the plant next to Clear River, which runs along the eastern edge of the town. Everyone likes to go there to catch the rainbow trout that live in the crystal waters. The senior citizen group has already made a proposal to build a housing project for senior citizens on the same property. They would guarantee that precautions would be taken to keep pollution from entering the river, and they would save the trees in the area by making the land around it into a park for the whole community to use.

The Chamber of Commerce is all for a new plant coming into town. They want to see the city grow, and new jobs would bring in new people as well as provide employment to the people of the community. The City Council agrees that new business would help the town because they would get new taxes for keeping city services up to date. They also want a new interstate highway to be built near the town. This would help in transporting goods and people. The community could really grow if this happens.

The P.O.E. (Protect Our Environment) group is angry. They want the community to know the risks involved if the plant moves in. Their river may have hot water and chemicals pouring into it. This will cause a terrible fish kill and end recreational fishing for the community. Tourists will no longer be attracted to their town for the scenery and for the fishing. They feel that the new jobs are not worth the damage to the community.

The Senior Citizens, which make up 40 percent of the population, feel that they can protect the environment and at the same time provide themselves with a safe place to live.

The firm does not guarantee anything, but they say they will try to be careful not to pollute, and they will pay their employees top dollar.

What should the community do?

Part 2
Countries of Our World

The world in which we live is made up of many nations: large and small, rich and poor, agricultural and industrialized, and ancient and new. In order to better understand them, we organize them geographically, culturally, politically, economically, historically, religiously, etc. The world can be seen as a quilt stitched together out of the nations of the world, creating a pattern of cultural and geographic maps that is interesting in its diversity and intricacy. Additionally, each piece of the quilt adds to the whole by being composed of different fabrics and weaves, different textures and colors that bring a richness of diversity which contributes to the whole global village in which we live.

Lesson 8
World Regions—An Introduction

Objectives

- To identify and examine the characteristics of a region
- To apply the regional characteristics to a problem-solving situation

Notes to the Teacher

In this lesson the definition of what constitutes a region is given and illustrated. It is important that students understand that regions are determined by some unique characteristic and not just by their physical descriptions. The first handout reviews some of these characteristics and how human interaction with the environment helps create some of these regions.

Sets of pictures obtained from the school or public library or from magazines that show different regions of the world and their distinctive features would be useful. After students read about regions and select pictures that illustrate regions, they complete a problem solving group activity. Students are given a situation with four site descriptions. Using the information they have about regions, students select a site and explain why that site was selected.

Procedure

1. Distribute copies of **Handout 22**. Read it aloud in class or assign it for homework. Visit the library and select pictures that reflect a variety of regions around the world (*e.g.*, deserts, mountains, swamps, lakes, people from various cultures and countries, examples of housing, industry, agricultural methods, etc.). Magazines like *National Geographic* are useful for this purpose.
2. Obtain from the library musical tapes from other cultures. The musical tapes shared reflect several types of music from other countries that demonstrate different musical styles from what students are used to hearing. Discuss music as it pertains to culture. Have students name other aspects of culture—food, dress, religion, form of government, language, customs, and traditions.
3. Explain that some regions of the world are based upon the culture of the people. Lapland, for instance, is defined by the area populated by the Lapp people. They have a distinct culture or way of life which sets them apart from others. The Lapland cultural region covers parts of northern Norway, Sweden, Finland, and Russia. Continue a discussion about the definition of region (areas of the earth that share some common or unique characteristics that can be physical, economic, political, or cultural).
4. Divide the class into small groups of four or five and, using **Handout 23**, have students decide which country would be best suited for having new industry move in.
5. Allow students enough time to discuss the costs and benefits of each choice. When they have made their choice and have prepared their list of reasons for that choice, reconvene the class and have students report on their decisions.
6. After determining which country got the most votes, reveal the names of the mystery countries.

   Suggested Responses:

   1. Netherlands

   2. Nepal

   3. Nigeria

   4. Nicaragua
7. Conclude by locating the mystery countries on a wall map. Continue to debrief the activity by discussing elements necessary to succeed in business in another country. (Some suggestions are stable government, educated/skilled workers, good relations between your country and theirs, good employer/employee relations, recognize cultural differences, natural resources available, and nearness to markets).

Enrichment/Extension

1. Working with one or two other students, research one of the mystery countries. Find out more about the kinds of things that make the region unique. Find pictures and maps of the country. Prepare a brief oral presentation for the class.

2. Use outline maps to show specific regions of the world. Prepare maps that show the desert regions, plains, coastal and interior, mountains, rainforests, etc. Create legends (in color or symbols) for each of these regions. Consider showing cultural regions based on language or religion. Use your maps to develop generalizations about conditions or events within a given region.

Name________________
Date________________

World Regions

Read this information and be able to describe the various ways to define a region.

What do we mean when we say we are going to study about world regions? Regions are areas of the earth that share common characteristics that can be physical (geographic), economic, political, or cultural. In this unit, world regions are going to be examined as political units (countries) from all seven continents. Their similarities and differences will be noted as we study these countries.

Some countries are highly developed with large populations who are involved in producing goods and services to supply the wants of their country as well as those wants and needs of other countries. Trade systems develop to allow those goods and services to be delivered to other nations who can pay for them.

Some countries are underdeveloped. In other words, these nations do not have the natural resources, the technology, or the financial backing to produce necessary goods and services. Often their populations are very large and untrained in the skills necessary to work as producers of goods and services.

Climatic and physical characteristics of a region determine the types of activities in which humans can engage in that region. Regions where the land surface is too dry or too mountainous, for example, are not good for agriculture. Humans do farm in some of these regions by irrigating and terracing the land, but their output is usually at a subsistence level (just enough for themselves) and a surplus that they could sell or trade to others is not likely.

Many regions have conflict with neighboring regions because of their cultural differences. When we speak of culture we are talking about the way in which people live—their languages, religions, customs, foods, dress, and form of government. Sometimes these conflicts keep countries from developing into powerful nations. Their cultural differences often cause resistance to change or to people from other countries who want to come in and develop their natural resources. Sometimes, boundaries between countries are in dispute, and a desire to expand their territories causes conflict with neighbors who may even share a common cultural background.

Understanding the unique characteristics of a region involves more than just reading a map. One must start with the physical geography, however, to understand the conditions in which humans are interacting with the earth. The physical and climatic conditions to which humans are subjected influence their cultural and economic development. A mountain range separating a region into two countries like Spain and France, for example, affected their cultural development. Their languages are different as well as their customs, traditions, foods, government, etc. If there had been no physical barrier between these two areas, it is possible that their cultures would have been more similar.

Regions do not always remain the same. Climate patterns change; new trading patterns develop; new uses for resources cause a change in the demand for certain natural resources; and a change in the economic development of some countries occurs. Our study of the countries in certain regions of the world will examine the conditions of those regions as they are today along with consideration of how those regions may interact in the future to create a different global picture.

Name________________
Date________________

Making the Right Choice

In small groups decide which country would be the best suited for having new industry locate there.

> Congratulations! Your company, Micro-Automobile Producers, has decided to establish a new factory in another country because it has been so profitable here. Your CEO has given your committee four possible choices. You must decide where that new location will be.

In your group, examine the countries described in the paragraphs. Using this information, decide whether or not your business would like to expand into this region. Make a list, for each country, of all the reasons for and against this proposal. After weighing these reasons, decide which of these countries would best suit your needs. Be prepared to share your decision with the class and the reasons for it.

Country 1

Population 15,368,000

Area 16,033 sq. mi.

Most of the land area is flat with the highest altitude at 37 feet above sea level. Much of the land is actually below sea level. The country has a parliamentary democracy, with a queen at its head. Its labor force has 6 percent in the agricultural area, 50 percent in the service area, and 16 percent work in the government. 89 percent of the population live in urban areas. There are a variety of industries located here along with a significant tourist trade. The literacy rate is 99 percent. This country is located in northern Europe.

Country 2

Population 21,042,000

Area 56,827 sq. mi.

In the northern part of this country, there is a very large mountain range with hills and valleys in the middle. Along the southern boundary, there is a subtropical plain. This country is in Asia. Its government is a constitutional monarchy with a king as its head. Most people work in the agricultural industry and tourism. Few cars are found here, but roads have been developed to some of the nearby countries. The literacy rate is 29 percent.

Country 3

Population 98,091,000

Area 356,669 sq. mi.

This country is about twice as big as California. It is located on a coast with one part of the country a large coastal mangrove swamp, a tropical rain forest, a plateau with grasslands across it and, in the north, a semidesert area. There are three large cities. Its industry relies heavily on its oil production and exportation. It has more oil deposits than any other African country. There are also food processing and automobile assembly plants. There are many privately owned cars. The government is not stable at this time. A military group is currently in charge. They are building a new capital city in the center of the country. The literacy rate is 51 percent.

Name________________
Date________________

Country 4

Population 4,097,000

Area 50,880 sq. mi.

This is a very small country about as big as the state of Iowa. It has coasts on both the Atlantic and Pacific Oceans. Down the center of the country are mountains with many volcanic peaks. There are two large lakes located east of the mountains. Their government is a republic. Most people (44 percent) are involved in agriculture with only 13 percent working in industries such as oil refining, food processing, chemical, and textile production. The remaining 43 percent work in service areas. They have only one television set for each twenty people. The literacy rate is 57 percent. This is a Central American country.

Lesson 9
Polar Ice—Where Are the People?

Objectives

- To locate in absolute and relative terms the polar regions
- To describe the polar regions' geographic features
- To identify the effects of human-environment interaction in the polar regions

Notes to the Teacher

This lesson takes students to the "ends of the earth" by locating the polar regions and examining the areas where people choose to live or not live. By studying lifestyles in these regions, students will notice how people adapt to these very cold environments. It is important to provide students with time and materials to do further research into the polar zones. Provide as much background as possible because students probably do not know a great deal about the polar regions.

Students read an introductory selection about the northern and southern polar regions. They complete a Venn diagram comparing the two regions. Map handouts are provided so that students can locate these polar regions. Finally, a small research assignment assists students in learning more about the people who live in the northern polar regions.

Procedure

1. Set a block of ice on a table for the class to examine. Ask them where they might be if that was their total environment. Then, discuss the polar regions to determine how much students already know. List on the board questions for which students would like to know answers about the polar regions.
2. After students have read **Handout 24**, discuss it. Have them complete **Handout 25**. Remind students how the Venn diagram is used with the intersecting areas for things the two regions have in common and the outside areas for things that are different. Once the Venn diagrams are completed and shared, have students check to see if they answered any questions that they wanted to know about before they began the assignment.
3. Distribute **Handout 26** so each student has maps of Antarctica and the world. Using student atlases, have students use a purple crayon or colored pencil and color all the tundra areas on the world map. With red pencil, draw in and label the Arctic and Antarctic Circles. Label Antarctica and the countries found in the tundra areas of the north. On the map of Antarctica, color the Ross Ice Shelf yellow. Circle the South Pole with Green. Color the water blue. On both maps, make hatchmarks in black to indicate uninhabited regions. Display students' completed work.
4. Divide the class into four cooperative groups and assign or have them select one of the groups on **Handout 27**. Students should either go to the library or use materials provided in the classroom to locate the information required. Remind students to illustrate as many of the items as possible. Encourage them to find pictures or copies of pictures that they could use as well. Have students put their research into written form. Have each group do an oral presentation to the class sharing the information gathered. Remind students that every group member should make a contribution toward the research and the oral presentation.
5. Do the presentations and display students' work in the classroom or hallway.

Enrichment/Extension

1. Find out about the marine life in Antarctica and do a mural with all the animals pictured on it. Present the mural along with background information about the animals to the class.
2. Find out about the animal life in the Arctic region and create a mural showing the animals. Present the mural to the class along with background information about the animals.
3. Research the hunting and fishing techniques of the Eskimo. Find or draw pictures of their weapons and present the report to the class.

Name________________
Date________________

Polar Ice—Where Are the People?

Read about the Polar Regions. Be ready to compare them.

BRRR!!! Is that what you think about when you hear the term polar ice caps? If so, then you are like almost everyone else in the world. Just think about the geography. This land lies above the Polar Zones—66½° N (Arctic Circle) and 66½° S (Antarctic Circle) and has the coldest temperatures on the earth. Yet, in some of these cold regions, people do choose to live and others choose to come and explore when the weather allows them to enter.

In the northern polar zones, one finds northern Canada, Greenland, northern Europe, Siberia, Alaska, and many small islands. In the southern polar zone, one finds Antarctica. Where are the people? They are not in Antarctica—at least not on a permanent basis. There are scientific stations run by the major world powers where scientists come and spend a year, but there are no villages or cities there—no permanent population. The climate is too severe.

Antarctica is the fifth largest of the seven continents. It is also the *coldest* continent. It is very windy and its inland plateau is classified as a desert because it gets only about two inches of precipitation in the form of snow each year. There are no trees or bushes. Most of it lies south of the Antarctic Circle. It is very nearly a circle in shape with a peninsula that stretches out and nearly reaches South America. In the winter, Antarctica nearly doubles in size because of the frozen seas around its borders. Seven countries have claimed Antarctica as their territory, but the Antarctic Treaty of 1961 set aside these claims and insists upon international cooperation in order to advance scientific research.

Many explorers have come to Antarctica since its discovery. One British explorer, Robert F. Scott, spent the years between 1901–1904 there gathering scientific information. He wanted to be the first person to find the South Pole and returned in 1910 determined to do so. He traveled by foot, refusing to use dog sleds. When he got to the South Pole, he discovered that a Norwegian explorer, Roald Amundsen, had already been there to raise the flag of Norway on the site. Disappointed, Scott started for home, but, due to a variety of miscalculations and weather problems, he ran out of fuel. He and the two men who were left on the expedition froze to death just a few miles from their base camp.

Richard E. Byrd, an American explorer, was the first aviator to fly over the South Pole along with three other pilots who were with him in 1929. (He and his copilot had made the first flight over the North Pole in 1926.) He made several expeditions to Antarctica and returned with new discoveries, scientific data, and maps of the region.

What about the North Pole region? This area is just ice—glaciers—but the tundras of Europe, Siberia, Alaska, Canada, and Greenland is land covered most of the year with ice caps with only a very short growing season in the summer when the tundra's layer of permafrost thaws down to a foot or two in depth. People and animals live in these regions adapting to the world's coldest climates.

Animals like the caribou or reindeer, musk-ox, wolf, Arctic fox, and polar bear adapt by developing thick waterproof hides and eating the life forms that are found in the region. There are a few birds like the ptarmigan and the Arctic owl that also live year-round on the tundra. Then in the

Name__________________
Date__________________

summer, large numbers of birds migrate to the region for the season. When the ice caps are permanent throughout the year, few warm-blooded animals can be found, but there are many marine animals like seals, whales, and walruses. There are no permanent human dwellers on the ice caps because there is no way for them to get food.

One group of people who live in the North American Arctic region are the Inuits—you may know them as Eskimos. They depend heavily on the marine life, especially the seals. Their homes are sod or stone houses that have been dug down into the ground. In the summer during the hunting season, they use caribou tents. In the winter, when they go out on the polar ice to catch seals, they build snow igloos—temporary housing for hunting trips.

The dogs they use for their sled teams are trained for this, but they are more like wild dogs in every other way. Families are small and people work cooperatively to get the necessary food, clothing, and shelter. Things are changing for the Eskimos in modern times. Modern style houses are being built and the conveniences that we know such as television, electrical appliances, etc., are being used in Eskimo homes. Their children want to look like other kids around the world so they are avoiding the traditional dress styles of their "old fashioned" families. The world is moving into the Arctic via television. Hopefully, Eskimos will share their ideas and culture with us—but, please, not the freezing temperatures. Brrr!!!

Name________________
Date________________

Polar Regions—Making Comparisons

On the Venn diagram below, list as many similarities between the north and south polar regions as you can find in the previous reading. Place them in the area where the two circles intersect. List the differences in the outer rings marked for each pole.

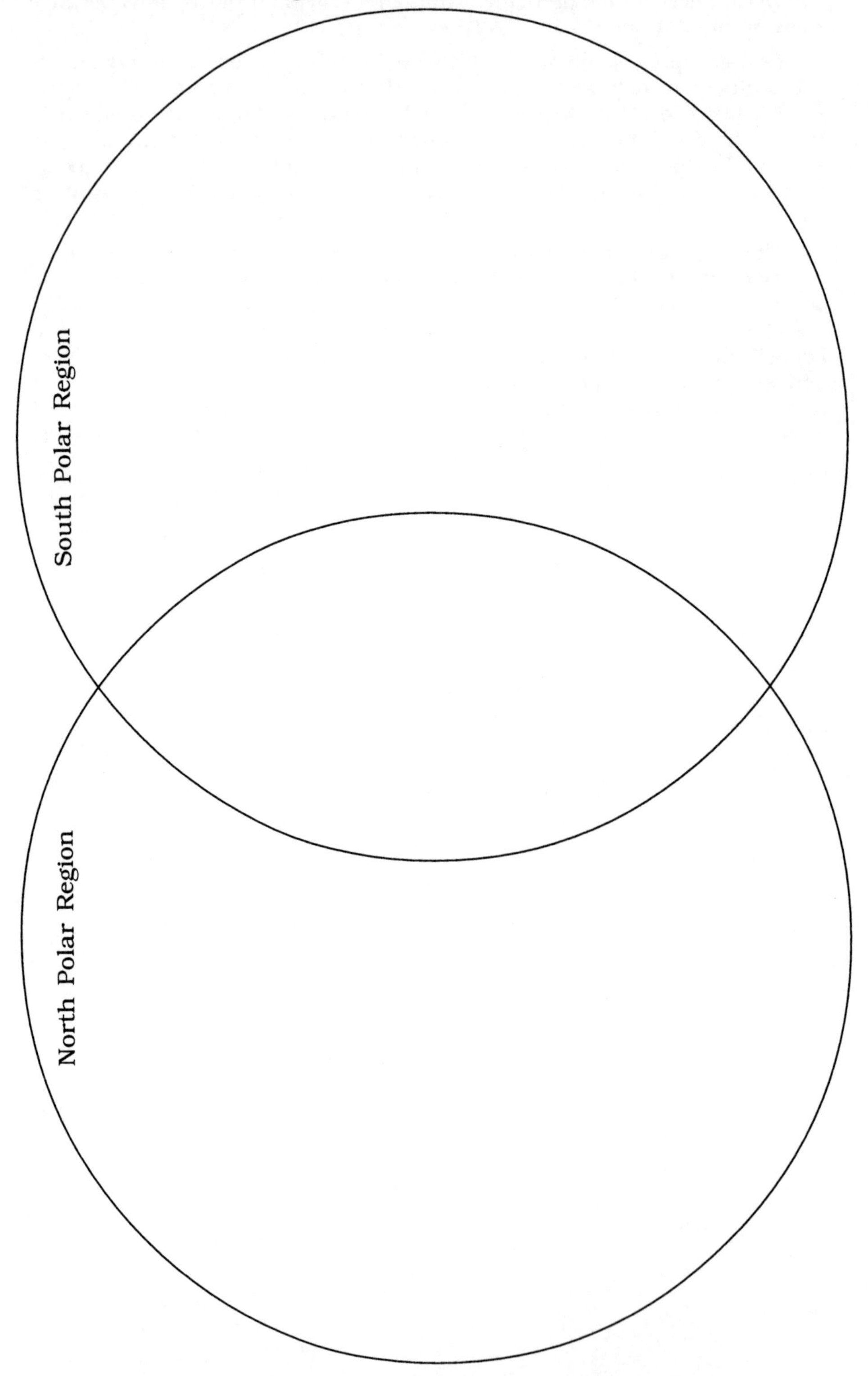

Name____________________
Date_____________________

The Geography of the Polar Region

Use an atlas to do the following:

On the World map

- Color all the tundra areas **purple**.
- Using **red**, draw in the Arctic and Antarctic Circles. Label them.
- Color the Ross Ice Shelf **yellow**.
- Label Antarctica and all the countries found in the tundra.
- Color the water **blue**.

On the map of Antarctica

- Circle the South pole with **green**.
- Color the water **blue**.

On both maps

- Make hatchmarks in **black** to indicate uninhabited regions.

Name____________________
Date____________________

World Map

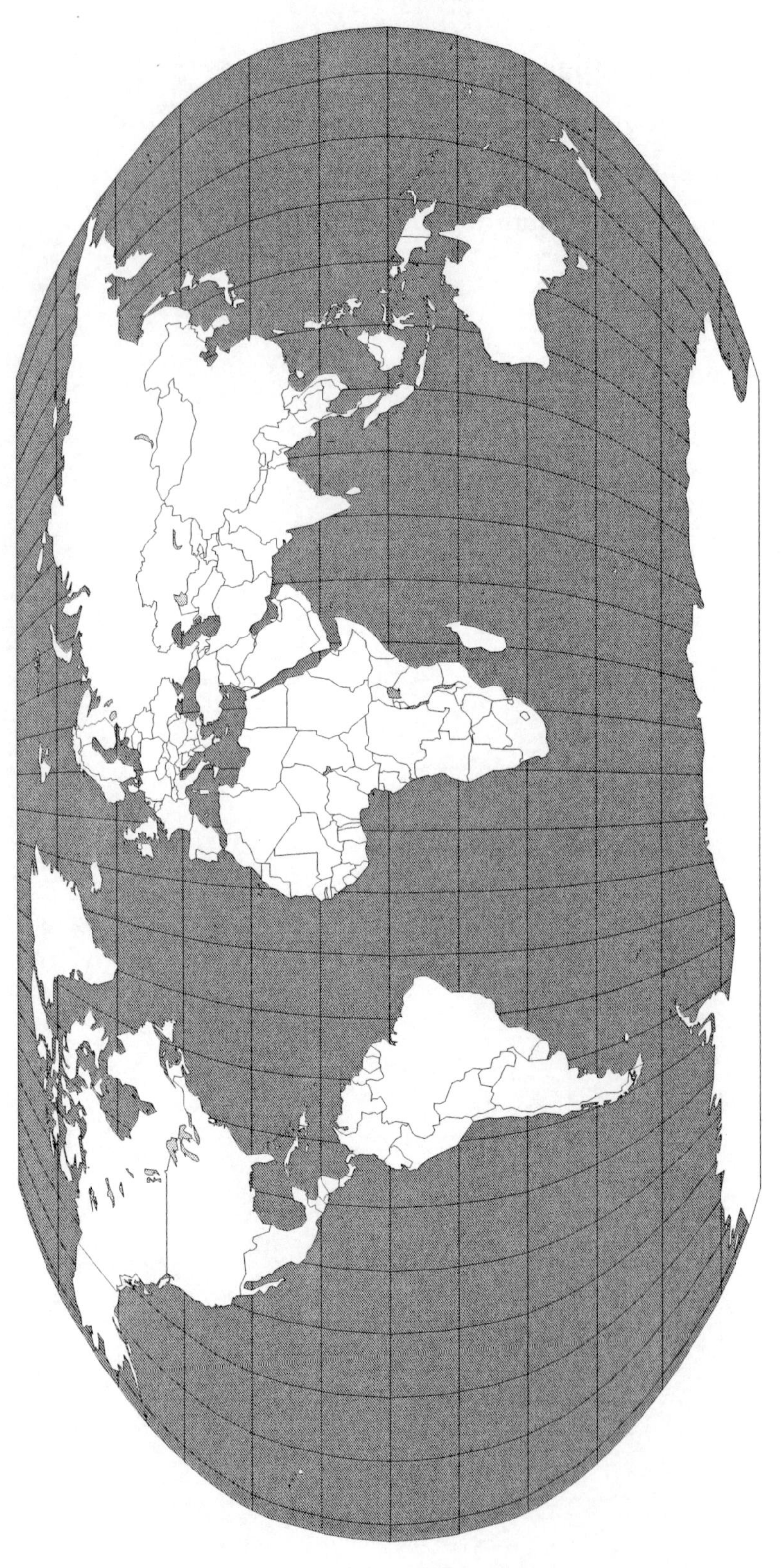

Name____________________
Date____________________

Antarctica

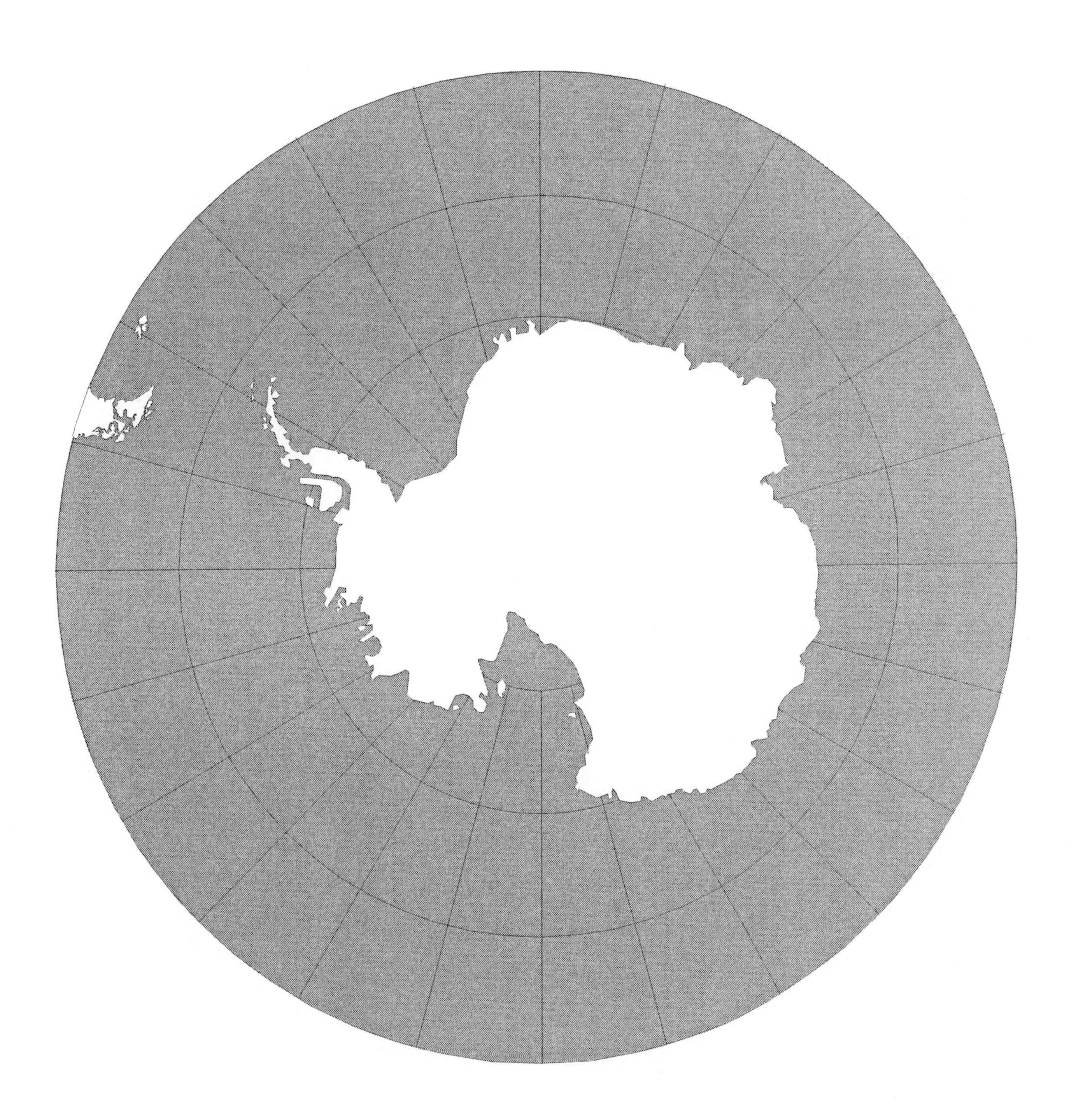

Name________________
Date________________

Polar Research

Listed below are four groups of people who live in the northern polar regions. Working with your assigned group, find out as much as you can about them.

Eskimos (Inuit)

Greenlanders

Laplanders

Icelanders

Things to know about these groups

a. kind of housing (illustrate the houses)

b. kind of clothing (illustrate the clothes)

c. kind of food

d. way they make their living

e. their natural environment

f. how they adapt to their environment

g. what they do for recreation

h. how they settle disputes

l. how they govern themselves

m. types of transportation available to them

n. how their lifestyles have changed in modern times

Lesson 10
Canada—Our "Cool" Neighbor to the North

Objectives

- To identify Canada's provinces and geographical features
- To examine the effects of climate and resources upon human settlement

Notes to the Teacher

Canada is America's northern neighbor. Most students do not know much about Canada. Since we share the world's largest and friendliest border, students should be familiar with the geography and cultures of the Canadians and the similarities between their origins and those of the United States. Gather as many materials about Canada as possible for students to use in the classroom or in the library.

This lesson involves students in some basic research and chart reading activities. Students will answer questions relative to the information provided in a chart. They also read a fictional journal about a young man's travels across Canada and relate it to the geography and settlement patterns of the area.

Procedure

1. Survey the class. How many have ever been to Canada? What do they know about this country? Let them talk for a few minutes about their impression of Canada based either on their personal experiences or on other sources.
2. Distribute **Handout 28** and remind students that most of Canada's population lives within a hundred miles of the U.S. border. Why would this be the case? (*Climate is too cold as one moves north.*) Have students locate and label the Rocky Mountains, Appalachian Mountains, the Great Lakes, the Canadian Shield, and the prairies. Which would be barriers? (*the mountains*) Which would allow freedom of movement? (*lakes, prairies, most of the Shield*) Where do you think people would choose to live? (*along lakes, prairies, and some of the Shield*) Why? (*easier access, better climate, easier to find work*)
3. Distribute **Handout 29**. Assign it as homework or complete as a classroom activity. Discuss the answers upon completion. Have students refer to their maps of Canada as each question is answered and locate the province or territory. Following the discussion, ask students why they think some provinces have more people than others. Have students look at their maps for assistance.

Suggested Responses:

1. *12*
2. *Quebec*
3. *Northwest Territories*
4. *Prince Edward Island*
5. *Ontario*
6. *Prince Edward Island*
7. *Yukon Territory*
8. *Quebec*
9. *French*
10. *Northwest Territories*
11. *Native American/Eskimo*
12. *New Brunswick, Nova Scotia, Ontario, Quebec*
13. *Northwest Territories*
14. *British Columbia, Alberta, Saskatchewan, Manitoba, Ontario, Quebec, New Brunswick*
15. *Yukon Territory*
16. *Ottawa, Ontario*

4. Help students examine some of Canada's larger cities by using **Handout 30**. Remind students that most Canadians live in the cities. What brings them to the cities? Why do people locate cities in certain areas? (location to natural resources, close to lakes or rivers for good transportation, location of a railroad or major highway–easy access, good water supply, good climate for growing things, etc.) Relate the growth of cities to the ability of humans to adapt to their environment.

5. On their maps, have students locate and label each of the cities where Sammy visited. Use an atlas and determine the number of miles between each leg of their journey and the total number of miles traveled.

6. Ask students to use library materials to find out more about each of these cities selected, including pictures of them. Working in groups, have students assemble a short report about each one and present to the class. Be sure the report compares the climate of the Canadian city the group researched to their own town. How are they the same? Different?

7. Have students write a paragraph about why they would or would not choose to live in the city they researched. Share paragraphs before polling the class to determine class consensus to the question.

Enrichment/Extension

1. Research other Canadian cities and create a travel journey of an imaginary trip across Canada indicating mileage, costs, famous landmarks, tourist attractions, etc.

2. Research how the Canadian government operates, indicating each branch and how it differs from that of the United States. Make a poster comparing the two governments and display it in your classroom.

3. Compile a list of famous Canadians. Select one famous Canadian and write a one-page biography.

Name__________________
Date___________________

Canada

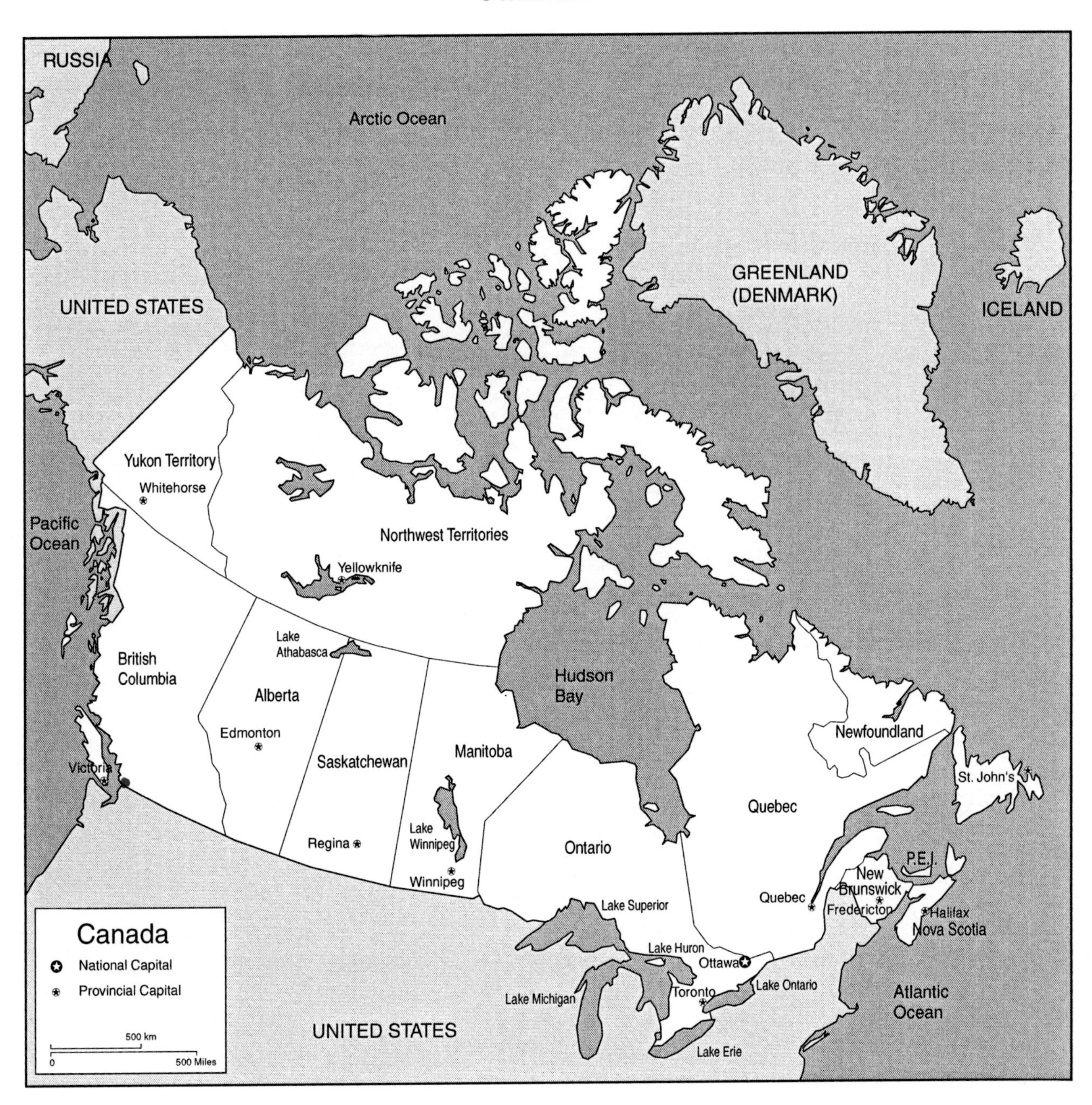

Name________________
Date________________

Drawing Conclusions From Facts

Answer the questions below using the information on page two of this handout.

1. What is the total number of provinces and territories in Canada?

2. Which is the largest province in size?

3. Which is the largest territory in size?

4. Which is the smallest province in size?

5. Which province has the largest population?

6. Which province has the smallest population?

7. Which territory has the smallest population?

8. Which province has a dominant language other than English?

9. What language is spoken there (question 8)?

10. Which territory has dominant languages other than English?

11. What language(s) are spoken there (question 10)?

12. Which four provinces were the first to become the Dominion of Canada?

13. Which was the last territory to join the Dominion of Canada?

14. Which seven provinces border the United States?

15. Which territory borders the United States?

16. Name the national capital and its province.

Name__________________
Date___________________

Canada—Fact Sheet

Province or territory	Capital	Size (sq. miles)	Population (c.1991)	Dominant language	Joined Canada (date)
Canada	Ottawa, Ontario	3,849,674	28,114,000	English	1867 Dominion established
Alberta	Edmonton	255,287	2,545,533	English	1905
British Columbia	Victoria	365,947	3,282,061	English	1871
Manitoba	Winnipeg	250,947	1,091,942	English	1870
New Brunswick	Fredericton	28,355	723,900	English	1867
Newfoundland	St. John's	156,949	568,474	English	1949
Northwest Territories	Yellowknife	1,322,909	57,659	Native American and Eskimo	1912
Nova Scotia	Halifax	21,425	899,942	English	1867
Ontario	Toronto	412,581	10,084,885	English	1867
Prince Edward Island	Charlottetown	2,185	129,765	English	1873
Quebec	Quebec City	594,860	6,895,963	French	1867
Saskatchewan	Regina	251,866	988,928	English	1905
Yukon Territory	Whitehorse	186,661	27,797	English	1898

Name_______________
Date_______________

The Journal of Sammy Smith

Read about Sammy Smith's trip to Canada. Be ready to discuss and map his trip.

Day 1 This has been an exciting day. I have packed my bags and I am ready to travel from my home in Ohio to see Canada. My parents told me that we will stop in some of Canada's important cities moving from the east coast to the west coast. I doubt if I will be able to sleep tonight.

Day 2 Dad made us get up at 4:00 A.M. to start our trip. I slept for the next three hours—then we stopped for breakfast. As we ate Dad told me some interesting things about Canada. Canada is the second largest country in the world but it has only about one-tenth of the population of the United States. Most of Canada's people moved there during the last 400 years. Only the Inuit and Native Americans lived there before that. The French were the first to settle Canada (1608—Quebec). Many Canadians still speak French today. The British took control of Canada in 1763. Canada became a nation in 1867—the Dominion of Canada. Mom said there are ten provinces in Canada and two territories.

Day 5 The geography of Canada determines where people live. People live in little pockets scattered around the country. Many people live around the Great Lakes and the St. Lawrence Lowlands, but the lakes and forests separate them from the people who live in Canada's prairies. The Appalachian Mountains separate the Great Lakes pocket from the Atlantic Coastal region where many people live. In the west, the Rocky Mountains separate the prairie people from those who live on the Pacific Coast. Canada must be really big to have so many pockets!

Day 6 Today we arrived in Quebec City, the capital of the province of Quebec and Canada's oldest city. It is strange to find so many people who do not speak English. I nearly got lost and no one could give me directions—except in French. Even the street signs were in French! Finally I met Jacques. He spoke English and French and he helped me get back to my parents. Jacques was a really cool police officer.

Day 7 I saw Jacques again today, but this time I wasn't lost! He told me that most of the people who live here work in government administration or tourist services. Every year they have a Winter Carnival, too. I would love to come back in the winter. Quebec is part of the Canadian Shield. Much of the land is covered with forests and is rich in minerals.

Day 8 The people who live in Quebec want to separate from the rest of Canada. That seems odd to me, and, besides, it did not work in the United States when our southern states tried to leave the union. Canada is different, though, it is a confederacy not a union. The rest of Canada does not want to see this happen. Maybe it has something to do with their being so French. French is the official language of Quebec while English and French are the official languages in the rest of Canada. Everything is handled in both languages—even street signs. Maybe I won't get lost in the next city!

Day 10 Well, we traveled on to Halifax, the capital of Nova Scotia and the largest city in the Maritime provinces. It has one of the world's largest harbors and the city is located on a peninsula. Halifax's harbor remains ice free, and this makes it very important to the military and to the Atlantic shipping business. Nova Scotia means New Scotland—I guess that tells you who settled this area. I bet those settlers liked it here because the weather is milder than it is over in the St. Lawrence Valley. They have cool summers and milder winters. There are many apple orchards and dairy farms. They also build ships. Over on Prince Edward Island, another of the Maritime provinces, they are famous for their potatoes.

Day 12 Now we are headed west again. We're going to stop in Ottawa, the capital of Canada. It is a beautiful city with the Ottawa River flowing through the center of it. There are many parks and in the summer, flowers are everywhere. I can

Name________________
Date________________

hardly wait to go to see the changing of the guard at the Parliament Building. The soldiers in their red coats and tall bearskin hats are wonderful on parade. They never smile—even when people make faces at them and do silly things! I know. I tried!

Day 13 I got to see the soldiers. They were great. We also toured the Parliament Building. I was impressed. Did you know that Canada has a prime minister instead of a president? Their legislature has two houses, the House of Commons and the Senate, which is the upper house. Their Senate is different than ours. They are appointed, not elected, to their jobs. Very interesting.

Day 14 We arrived in Toronto today. It is the capital of Ontario. Today there are many industries in Toronto. Dad says Toronto is important because it was a port city, and industries and people came here making it a trade center. They make machinery and electrical equipment, furniture, rubber and plastic products, and clothing. Also, food and beverages are packaged for distribution and printing and publishing are done here. The BEST thing about Toronto is the CN Tower. It is the world's tallest free-standing structure. The glass elevator took me up to the observation deck—more than a 1000 feet high. It was cool! I was there almost before I knew it. I could see all of Toronto from the Tower.

Day 17 We got to Edmonton, the capital of Alberta, today. I am really tired. We have driven many hours to get to Edmonton. I hope to go exploring tomorrow.

Day 18 Today we drove around Edmonton. The city is very busy. The north Saskatchewan River flows through the city. Oil was found there almost fifty years ago. There are lots of oil refineries now. They also have many food processing industries. It is a major commercial area, and many people work in government offices. Alberta is one of the prairie provinces.

Day 23 We have continued across the prairie provinces for several days. In the southern region I saw lots of cattle and wheat fields. If we went north we would find ourselves in the territories, but we do not have time to explore these cold regions. It's a good thing that I packed a jacket, though, because the nights on the prairies are quite cool. If we went farther north I would need even warmer clothes.

Day 26 Yesterday we crossed the Rockies. The mountains are so much more rugged than the Appalachians back east. They really are rocky. We will arrive in Vancouver soon. I can hardly wait.

Day 28 Well, after a couple of wrong turns, Dad got us to Vancouver. Mom said we could have saved two hours if he had just stopped to ask directions when we came off the interstate. Vancouver is Canada's third largest city. It is in the province of British Columbia. It was settled when the Canadian Pacific Railways located its transcontinental terminal here. It had a good harbor and, with the railroad here, people soon came to build their businesses. They ship many goods to countries in the Pacific. Shipments of wheat, lumber, potash, and coal are sent overseas. They also have wood, furniture, paper, publishing, and food and beverage industries. Vancouver is a major financial center. I like it. They have everything. I can go to the ocean to swim or to the mountains to ski or stay right in town and picnic in one of the parks. Tomorrow, I want to go to Chinatown.

Day 30 Today we have to start back home. Mom and Dad said we could stop at some of the small villages along the way since we had seen most of Canada's large cities already. Seventy-five percent of Canadians live and work in the cities. Few people live in the Northern Territories because of the severe climate. However, the Inuits and Native Americans have learned how to live in these cold climates. I am sorry that we could not go to visit there. Maybe next time.

Lesson 11
Mexico—Beyond the Rio Grande

Objectives

- To investigate Mexico's history and present
- To locate geographic areas of the United States that have direct historical and cultural links to Mexico
- To demonstrate human-environment adjustments to geographic situations

Notes to the Teacher

The United States and Mexico are connected by many historic and cultural ties. The two countries are linked by the exploration and settlement patterns of Europeans coming to the New World in the sixteenth century. The histories of the two countries have intertwined often over the last five centuries.

Currently, Mexico is facing a set of difficult problems, and the relationship between the United States and Mexico reflects the impact of those problems (*e.g.*, the number of illegal Mexican immigrants who flee to America to escape the poverty of their homeland).

Additional lessons concerning Mexico can be found in other Center for Learning publications. The biographies (Montezuma, Maximilian, Rivera) in *International Biographies, Book 6—Latin America,* feature persons of significance in Mexican history.

Students read a selection that provides background information on Mexico. They color states on a map to show where the U.S. boundaries were expanded to include former Mexican territory and to figure the percentages of the United States that belonged to Mexico at those points in time.

Procedure

1. Greet the class in Spanish. (Buenos Dias) Ask them if they know that phrase and what it means. (Good Day) Determine if there are any Spanish speakers in the class. Ask them to teach the class some other phrases. Discuss why there are people in the class or in the United States that speak different languages. Ask them to speculate on how those people came to be in the United States.
2. Have students read **Handout 31** and answer the questions that follow. Make sure that students have an opportunity for discussion. Consider forming student groups of three or four to answer the questions cooperatively. A large wall map of Mexico is helpful.

Suggested Responses:

1. a. *Many Mexicans stayed in the United States after the territory changed hands. Others migrated to be with friends and family and to find work.*

   b. *Spanish can be heard frequently; there are Mexican restaurants; many workers are Mexican-Americans; the schools offer Spanish more often as a language to study, etc.*

2. a. *The Mexican-American War*

   b. *No, many stayed because the area was their home and they didn't want to leave.*

3. a. *The rural farmers (often Native Americans), middle class in industrial/business areas, a transitional group trying to move into urban areas by living in nearby shantytowns*

   b. *People live in shantytowns hoping to find work and to be able to move up the social ladder.*

4. a. *The Aztec Empire and gold*

   b. *The Spanish changed everything—forced the natives to work for them and tried to change native languages, religion, and customs.*

5. *The altitude influences crops grown. Different crops are grown at different altitudes because the climate is different as the altitude rises.*

6. *Extreme poverty has affected the people by causing some to move into shantytowns hoping for better jobs; some enter the United States illegally to seek employment; the government is not as stable because people want better economic conditions.*

7. *Los Angeles, California*

8. *a. Oil*

   b. The price of oil fluctuates and, so far, their oil has not had as many financial benefits as they had hoped.

3. Have students work in small groups to determine how Mexico's problems might be solved so that its citizens could remain at home and have work in their own country. Be sure to remind students that Mexico's population is growing at a very rapid rate, and that in the future there will be even more people looking for work than there are today. Provide time to share and discuss groups' suggested solutions.

4. Have students use **Handout 32** to color in the various states that represent areas which formerly belonged to Mexico. To figure the percentages, students need to place the number of states shaded over the total of fifty states. By dividing the denominator into the numerator, a percentage is obtained.

   Suggested Responses:

   1. 14% (7/50)

   2. 4% (2/50)

   3. 28% (14/50)

   4. 46% (23/50)

5. Ask students if they are surprised to find such a large Mexican influence in the United States. Ask how what they have learned has caused them to reconsider Mexico as a neighbor.

6. Display completed maps.

Enrichment/Extension

1. Use a large piece of poster board paper to make a map of Mexico. Color the map to show physical characteristics. Cut the map into a puzzle based on the states' boundaries. Review Mexico's geographic features and political boundaries by practicing/reassembling the puzzle.

2. Create a bulletin board collage of Mexico using maps, pictures, quotations, etc.

3. Create a Mexican cookbook by collecting recipes at the library. Prepare a Mexican feast for classmates or eat at a Mexican restaurant.

Name________________
Date________________

Mexico—Southern Neighbor to the United States

Read the selection below and answer the questions which follow. Be prepared for class discussion.

Like Canada, Mexico shares many historical, geographical, and cultural ties with the United States. Spanish explorers visited or explored twenty-three U.S. states, made land claims in fifteen of them, and established settlements in seven states. California, Utah, New Mexico, Arizona, Nevada, Texas, and Colorado were all once part of Mexico. After the Mexican-American War, the United States acquired almost half of Mexico's territory. In California and the southwestern states, many people speak Spanish, and Mexican culture and heritage is widespread.

The United Mexican States is made up of thirty-one states and one federal district. The population is 92,202,000 with thirty-eight percent of the people under fifteen years of age. Sixty percent of the population is Mestizo (Spanish and Indian), thirty percent is American Indian, and nine percent is Caucasian. Seventy-one percent of the population live in urban areas. Eighty-nine percent of the population is Roman Catholic. The government is a Federal Republic with a president as the head of state.

The people of Mexico can be divided socially into three groups: the rural farmers, often Native Americans who follow the old, traditional lifestyles; the urban upper and middle class, who live lifestyles related to the business and industrial world; and a transitional group trying to move into the urban societies while living in shantytowns around and in the large urban areas like Mexico City, the capital.

Mexico has a large central plateau. The northern part of the plateau is too dry for farming, but the southern part is more productive and contains almost half of Mexico's population. Less then twelve percent of Mexico's land is fertile but more than forty percent of the population work on farms. The most productive area lies in the valley of Mexico, the location of Mexico City. Since the altitudes are high in this region, different crops are grown as farmers work at different distances above sea level. In the low, tropical areas, tobacco, sugarcane, sisal, coffee, and bananas are grown. Between 3,000 and 8,000 feet in the more temperate climates, maquey (source of tequila), corn, wheat, and beans are grown. At altitudes above 8,000 or 9,000 feet it is too cold for farming. Few people live beyond this altitude.

Mountains lie to the east, west, and south of the Central Plateau with the eastern and western mountains meeting a little south of Mexico City. Many huge volcanic cones exist. Along the coastlines are plains areas. The eastern coastal plain is semiarid in the north with wet, jungle area in the south. The western coastal plain is narrow and has not seen much economic development. There is the Sonora Desert in the north and a mountainous region in the Baja California area. In the south, there is swampland and jungle. The heat and rainfall allow for the production of sugar cane, bananas, and coconuts. The sap from the sapodilla tree is gathered here for making chicle, an important ingredient in chewing gum.

Mexico's culture extends far into the past with the Toltec and Aztec Indian civilizations. The Spanish found the Aztecs and their extensive empire in place when they arrived in the sixteenth century. The Spanish defeated the Aztecs, colonized the land, and changed their culture by introducing Hispanic language, religion, and customs. Evidence of this change is seen today with Spanish as the official language, but the American Indian languages are still spoken by many people. Many buildings in Mexico reflect the Spanish architectural style.

Name________________
Date________________

Mexico is trying to compete in world markets with industries working in these areas: steel, chemicals, electric goods, textiles, rubber, petroleum, and tourism. The oil industry has had help from foreign investors, but because the price of oil goes up and down, it had not earned as much money as the government had hoped. Many people in Mexico are jobless. Programs have been set up to help the poor, but there is political unrest. Presidential candidate, Luis Donaldo Colosio Murrieta, was assassinated at a political rally in Tijuana on March 23, 1994.

There is a link between the United States and Mexico culturally. Los Angeles has a large Mexican population—surpassed only by Mexico City. Many Mexicans try to enter the United States each year illegally. It is a problem for people living along the border areas because they feel that illegals take jobs from Americans. The U.S. Government has taken measures to keep illegal Mexicans at home, but they are still coming because of the poverty found in Mexico. The Mexican government is trying to bring about changes in the economy and to improve things there, but this takes time and money. While Mexicans want to do it themselves, they often have to rely on outside investors.

Name________________
Date________________

Questions to consider

1. a. Why are there so many Mexican-Americans in the Southwest and in California?

   b. What influences are evident in cities where they live?

2. a. What event occurred in American history that changed the boundaries of Mexico?

   b. Do you think that people left their homes just because they were no longer part of Mexico? Why?

3. a. What are the three social groups found in Mexico today?

   b. Why do you think so many people have come to the cities to live in shantytowns?

4. a. What did the Spanish find in Mexico when they arrived?

   b. How did the Aztec lifestyle changed after the Spanish overran their empire?

5. What influences the type of crops grown in Mexico? Explain.

6 What effect has the extreme poverty had on the people and the government of Mexico?

7. What U.S. city has more Mexicans than any other city except for Mexico City?

8. a. What mineral has been discovered in Mexico that many believe will help ease their bad financial situation?

   b. Why hasn't it helped as much as the government had hoped?

Name________________
Date________________

Mexican Influences on the United States—Historical Perspectives

Part A.

To complete this handout, you will need three colored pencils, crayons, or markers: blue, green, and yellow. You must study the map in part B of the handout to complete your coloring (explained below) on part C of the handout.

1. With your **blue** crayon or marker, color the following states which were part of Mexico prior to the Texas Annexation of 1845 and the Mexican Acquisition of 1848: California, Nevada, Utah, Arizona, southern Colorado, New Mexico, and Texas.

   Question: What percentage of the United States was once part of Mexico?

2. With your **green** crayon or marker, color the following states and territories which were at one time part of New Spain (modern Mexico): Florida and Puerto Rico.

   Question: What percentage of the United States was once a part of Spain's Mexican empire?

3. With your **yellow** crayon or marker, color the following states which were either first visited, explored or settled by Mexico (as New Spain): Alabama, Arkansas, Georgia, Kansas, Kentucky, Louisiana, Mississippi, Nebraska, North Carolina, Oklahoma, Oregon, South Carolina, Tennessee, and Washington.

   Questions: What percentage of the United States was either first visited, explored, or settled through Spain's Mexican empire?

4. Determine the total percentage of the United States that has been influenced by Mexico by adding up all the percentages from questions 1, 2, and 3 (total percentages).

Name________________
Date________________

Part B.

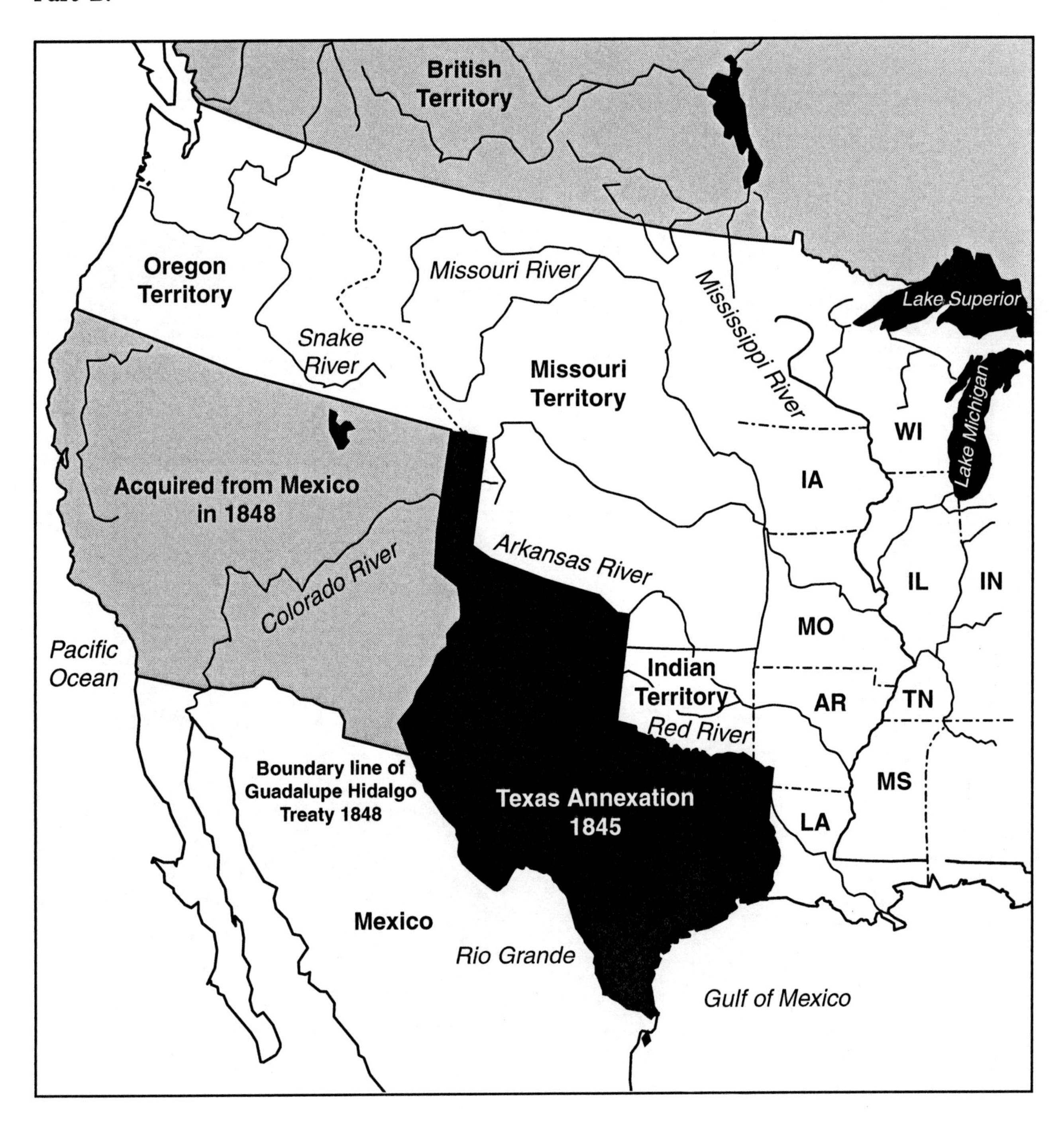

Name__________________
Date___________________

Part C.

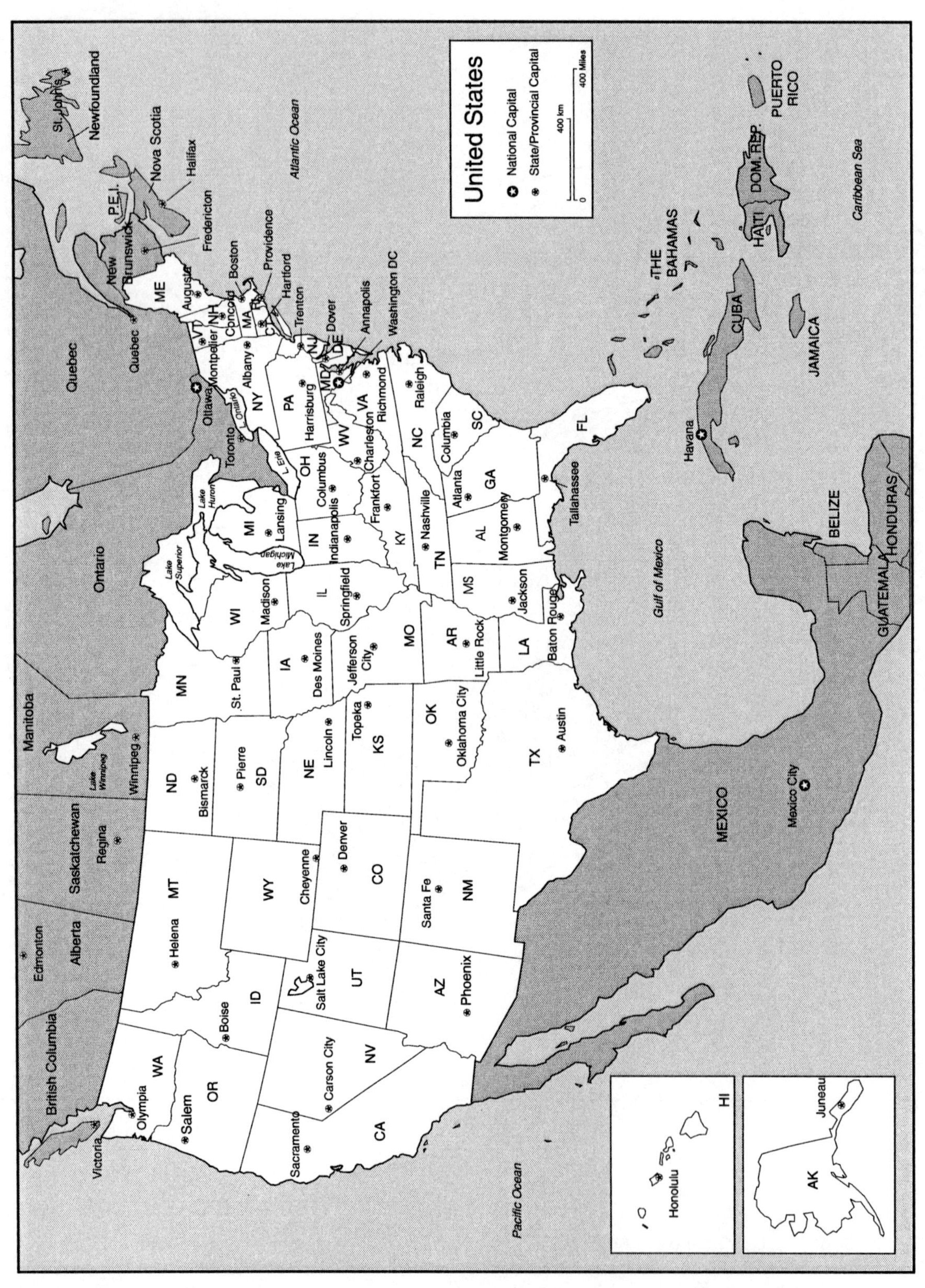

Lesson 12
Panama—Linking the Oceans

Objectives

- To introduce Panama as a representative Central American country
- To offer opportunities to read for information
- To offer opportunities to translate information into modes that demonstrate learning

Notes to the Teacher

To most students, Panama and the Panama Canal are synonymous. This lesson draws on that student preconception by providing a reading on Panama and its canal which directs students to read for specific information, some of which becomes the basis for the activities that follow.

Since Panama is an isthmus, and because of its location, it became a logical location for a canal. By stressing this point, students should be able to see how a distinctive landmass, such as an isthmus, shapes, or can shape, the course of a country.

The activities provided can be presented as individual or group activities, as selected activities, or as a sequence. These options provide the teacher materials that can be matched to the abilities of the students. The completed activities can be used for bulletin board displays to reinforce student learning visually.

Procedure

1. Begin by asking students what they know about Panama, information that draws from either textual reading assignments or from students' knowledge of geography. (It is an isthmus; it has a canal; it is the transition from Central to South America.) Ask students to give reasons why Panama was a logical location for building a canal (an isthmus, located between continents and oceans). Help students realize that landmasses may shape the course of a country.
2. Distribute **Handout 33**. Allow students time to read the selection and to answer the questions. Have students share their answers so they can verify that they located the correct information in the reading.

Suggested Responses:

1. *A narrow strip of land between two large bodies of land*
2. *The Atlantic and the Pacific; North America and South America*
3. *The economy is based more on international trade than on agriculture.*
4. *Mixed races*
5. *Spanish in 1520s*
6. *France in 1881*
7. *United States in 1914*
8. *1999*

3. Have students select one, several, or all of these activities.
   a. Make a bar graph or a pie graph showing the ethnic population make-up of Panama.
   b. Trace a map of Panama and color in the area, with clear outlines, that shows where the Canal is located.
   c. Create a chronological timetable of the Panama Canal from the 1520s through 1999. Use the handout, textbook, and/or library research.
   d. Suppose that for five miles on each side of the Mississippi River, Panama was the government in control. People could be stopped by Panamanian police, tried in Panamanian courts, could not live in the Mississippi River Zone, yet could work there—but for less money than a Panamanian doing the same work. Write one paragraph explaining how this situation would make you feel. Write a second paragraph explaining how this has helped you to understand why Panama wanted full control of the Canal Zone in 1999.

Enrichment/Extension

Prepare a short (five minute) oral report or a short (two page) written report on one of the following canals:

- Houston, Texas Canal—U.S.A.
- Erie Canal—U.S.A.
- Suez Canal—Egypt
- White Sea-Baltic Canal—Russia
- North Sea Canal—Germany and Denmark
- Manchester Ship Canal—England
- Welland Canal—Canada
- Grand Canal—China

Locate and include in your report comparable information to that with which you have studied on the Panama Canal: length of canal, dates of building, time spent building, number of locks, who has control, and landmass features related to the canal.

Name__________________
Date__________________

Panama—Crossroads of the Continents

Read the selection below and then answer the questions that follow.

Panama is located on a geographic landmass called an isthmus, that is, a narrow strip of land connecting two large bodies of land. Panama links the continents of North America and South America and, through the Canal, links the two largest oceans, the Atlantic and the Pacific. Because of the location of both Panama and the Canal, it has become the international banking center of Latin America. Panama City alone has over 100 banks. Panama is the only Central American country that has an economy based more on international trade than on agriculture, giving the people of Panama a higher personal annual income than their neighbors.

The people of Panama are comprised of several groups. Eleven percent are Spanish, thirteen percent are Black, sixty-five percent are mixed race, and ten percent are Indian (Guaymi, Cuna, and Choco tribes). There are also thousands of American citizens who reside either in the Canal Zone or on American military bases.

The idea of a canal that would link the oceans was first suggested by Spanish colonists as early as 1520. This idea was almost realized when, in 1881, France attempted to build a canal. However, both technical problems and disease halted the Canal when it was only one-third completed. In 1903, the United States purchased the Canal rights for ten million dollars and completed the remaining two-thirds of the Canal in 1914. The Canal was under American jurisdiction until 1979, when Panama assumed partial control in a transition that granted complete control in 1999. As a result of this transition, 80 percent of the Canal workers currently are Panamanians.

Fifty miles long, the Canal uses six locks to raise and lower the water level eighty-five feet in order to move ships from one ocean to the other. Almost half of the Canal cargo is United States in origin, and 26 percent of the cargo goes to U.S. ports.

1. What is an isthmus?

2. Panama joins which two oceans? Which two continents?

3. Why do Panamanians have a higher personal income than their neighbors?

4. What group makes up the largest ethnic group in Panama?

Name________________
Date________________

5. Who first had the idea for a canal? When?

6. Who began construction on the Canal? When?

7. Who completed the Canal? When?

8. When did Panama gain full control over the Canal?

Lesson 13
Brazil—Finding Its Place in the World

Objectives

- To introduce Brazil and its significance in the world
- To consider how a geographic feature such as a river basin helps to shape a culture

Notes to the Teacher

If Panama is the Canal, and Switzerland is the Alps, then Brazil is the Amazon. The primary focus of this lesson is that students gain an understanding of how a geographic feature—a river basin—influences the development of a nation or culture.

Brazil is perhaps South America's most significant country. Its three million square miles cover nearly half of South America, and its 113 million people represent half of South America's population, which means that half of South America speaks Portuguese, the national language of Brazil. The fifth largest country in the world, Brazil has the world's largest river basin, the Amazon, and one of the world's largest forests—20,000 varieties of trees! Completed in the 1990s, the Itaipu power generator—costing six billion dollars—is the world's largest.

Two of the fastest growing cities in the world are in Brazil. Rio de Janeiro, at 9.6 million, adds another million population every three years. Sao Paulo is the largest city in the Americas and the second fastest growing city in the world (after Mexico City). The population is nearing twenty million people. Both of these industrial cities are on the East Coast.

Yet, outside of these two larger cities, it is the Amazon which has shaped, and will continue to shape, the culture of Brazil. The first handout in this lesson has students speculate on how this river has helped to shape Brazil. Students then read and discuss a reading on the importance of rain forests.The final handout asks students to create an illustrated map demonstrating Brazil's economic potential.

Students will need access to almanacs, encyclopedias, and other references on Brazil.

Procedure

1. Begin by sharing information from the Notes to the Teacher with students. One way to generate student interest is by asking questions:

   What is the largest city in the Americas? (*Sao Paulo*)

   What is the largest river basin in the world? (*Amazon*)

   Have students read their text's material on Brazil.

2. **Handout 34** is designed for students to consider how a geographic feature like a river basin can shape a culture. This can be done by students either individually or in small groups. If done in groups, each group may select a leader to report responses to the class. Then, one student can summarize, on the chalkboard, the results of the groups. Distribute the handout, allowing ample time for students to read the questions and record their responses. Have each student complete the handout questions whether or not this is a group activity.

   Suggested Responses:

   1. a. The river would provide transportation to and from the interior; would provide water to drink.

   b. Because of transportation and water, villages, cities, or ports would be built along the river.

   c. At first, small boats; later, larger barges and transport ships will carry goods, information, and travelers.

   d. Water can be used for irrigation along the river; silt deposits will provide for good farmland.

   e. Fish would be part of the diet; other food would be limited to what is grown along the river or can be transported along the river.

   f. *Essentially, the river would become a central communications network (e.g., mail boats; information carried by passengers).*

   g. *At first, where possible, water mills; potential for dams and electric generators.*

2. *Travelers, consumer goods, equipment for further development*

3. *Timber, raw materials, people*

4. *Students should have arrived at the conclusion that a river becomes a transportation, communication, and development network for a nation.*

3. Distribute **Handout 35** and read it aloud. Depending on the reading level of the students, a volunteer might be selected to read it to the class. Discuss the importance of preserving rain forests. Inform students that they will use this information in the next handout.

4. Distribute **Handout 36** and have each student complete the map activity. Consider having students work in groups to compile the required information. (Note: It is important to have almanacs, encyclopedias, and other references on Brazil available for this aspect of the lesson.)

Enrichment/Extension

1. Research the plants of the Brazilian rain forest and their importance to modern medicine.

2. Investigate the connection between the rain forest and the greenhouse effect and the dangers of global warming.

3. Prepare reports on the political and economic problems—unstable government, inflation, gaps between rich and poor, and widespread poverty—that prevent Brazil from being prosperous and progress from being made in finding solutions.

Name________________
Date________________

The Amazon River Basin

Considering that the Amazon River basin is the central geographic feature in Brazil, think about how it has shaped the development, economy, and culture of Brazil. Keeping this in mind, answer the questions. Refer to the map to stimulate ideas.

1. How do you think the Amazon has influenced, or will influence, the following:

   a. exploration

   b. location of villages and cities

   c. transportation systems

   d. farming and agriculture

   e. food and diet

   f. communications

   g. energy development

   h. other

2. What goods would go down river to the sea?

3. What goods would go up the river?

4. What has this handout helped you to understand about the way that rivers, as a geographic feature, can shape a culture or a nation? Cite specific examples.

Name________________
Date________________

Why Preserve the Forest?

Examine the following.

Rain forests are intimately tied to global weather conditions, and their preservation helps prevent the global warming trend known as the Greenhouse Effect. Deforestation causes up to 30 percent of all human-produced carbon dioxide to the atmosphere, as well as unknown amounts of methane and nitrous oxide—gases that exacerbate global warming and threaten the quality of life worldwide. Rain forests also provide a natural defense against hurricanes, cyclones and typhoons, absorbing the punch of howling winds and preventing storm tides from eroding beaches.

The products that come from the rain forest are part and parcel of our daily lives. As a fantastic natural pharmacy the forest is home to thousands of medicinal plants that can be turned into antibiotics, painkillers, heart drugs and hormones. The National Cancer Center in the U.S. has identified 3090 plants as having anti-cancer properties—70 percent of which come from the rain forest. Other products that can be taken from the forest without destroying it range from cosmetics to automobile tires.

Tropical forests also provide the planet with much of its biological diversity. Every species that lives there is a living repository of genetic information, *i.e.*, the building blocks of life. If the food chain that binds then together in a complex web of relationships is disturbed, it is not clear that humankind itself could survive. At the very least, we'd be facing the future with a shrunken world, a hostile climate, and a genetic base vulnerable to mutations.

Last but not least, the forest is home to millions of indigenous peoples, who have known no other way of life for thousands of years. Within their memory banks is a trove of natural wisdom, including how to use plants medicinally, that can never be reduplicated once they pass from the earth. With very few exceptions, the forced relocation of indigenous forest people in the face of the bulldozer has invariably spelled disease, despair and death.[1]

[1]Pamela Bloom, *Amazon Up Close* (Hobe Sound, Fla.: Hunter Publishing, 1997).

Name________________
Date________________

Brazil—Unrealized Promise

Brazil, the fifth largest country in the world, has enormous possibilities for growth. Its major river, the Amazon, pours more water into the sea than the next seven biggest rivers in the world. Its volume is ten times that of the Mississippi River and at least fifty times that of the Nile River. The basin of the river is nearly as large as the continental United States. The Brazilian rain forest offers numerous benefits to the world. In addition, Brazil contributes many natural resources, crops, and manufactured items to other countries. Unfortunately, however, problems of widespread poverty, unstable governments, runaway inflation, and huge gaps between rich and poor have prevented it from reaching its potential.

To conclude this lesson, complete the accompanying visual of Brazil's economic promise by following these steps:

1. Color the rain forest of Brazil **green**. It covers the states of Amazonas, Acre, Rondonia, and most of Para, Mato Grosso, and Maranhao.
2. Draw in and label the Amazon River.
3. Locate and label the capital city of Brazilia.
4. Locate and label the cities of Sao Paulo and Rio de Janiero. Use dots to indicate the precise location of these cities.
5. In the upper left corner of the page, make a list of major agricultural products of Brazil.
6. In the upper right corner, list four benefits of the rain forest.
7. In the lower left corner, list major industrial products of the country.
8. In the lower right corner, list major natural resources found in Brazil.
9. Complete the visual by researching the information required along the outer border of the visual.

Your textbook, encyclopedia, and world almanac are useful resources in completing this assignment.

Name________________
Date________________

Agriculture

Rain Forest

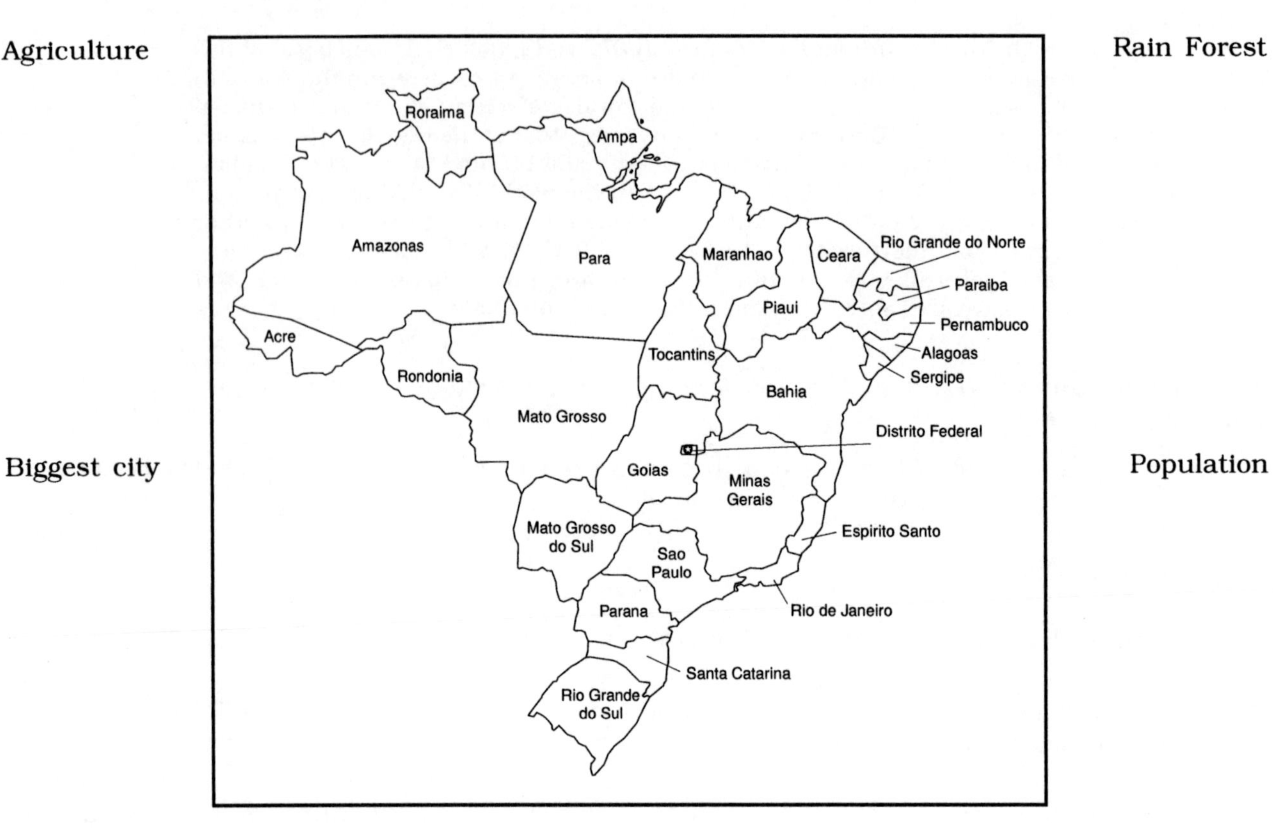

Biggest city

Population

Biggest port

Rate of literacy

Industry

Percent of natural increase in population

Natural resources

Lesson 14
Peru—Rich Man, Poor Man . . .

Objectives

- To identify and analyze some of Peru's economic problems
- To examine the cause and effect of political unrest in Peru
- To relate the geographic themes of location, place, regions, and human-environment interaction with the development of Peru

Notes to the Teacher

Students should understand that all is not well in the "Land of the Incas," as Peru has been historically known. The political situation is one of instability. It reflects the economic unrest generated by the many poor people who resent the small, wealthy landowner class. Students may need to be reminded that when the Spanish came to Peru in the sixteenth century they found a highly successful empire run by the Incas. However, the Inca Empire had been weakened by civil war between two brothers whose father had been the last ruler. The Spanish took over, carried away all the valuable gold, and imposed their rule over the Incas. Evidence of this Spanish reign is still present in the Spanish language, architecture, religion, etc., of the area. (For further information about the Incas and their last ruler, Atahualpa, see the Center for Learning publication, *International Biographies: Latin America*, Lesson 2.)

After reading about Peru and discussing the reading, students mark a map of South America with specific sites. Help students make generalizations about the location of Peru in relation to the equator, the countries around it, and the regions in and around it. Students discuss why Peru's population is in certain areas, and how people might be encouraged to move into unpopulated areas with economic incentives. Students write a help-wanted ad to entice people to take jobs in some of the areas where the government would like to see more people settle. Note that students will need atlases and colored pencils to complete this lesson.

Procedure

1. Ask students if they like mashed potatoes. french fries? baked potatoes? boiled potatoes? Use tally marks and chart their responses on the board. Have students create a bar graph showing their "potato preferences." Then, ask students if they know in which country the potato was first grown. You may wish to tell them Peru (grown first by the Incas) or you may start this at the end of class and tell them to find out for homework. Discuss other information that they may know about the Incas or Peru.

2. Assign **Handout 37** as homework. Discuss the reading with students. Emphasize the class differences and economic wealth that separates the classes. Discuss how the class would feel if the teacher gave most of the supplies to a small group of students and the rest had to do the best they could with very little or no supplies. Ask them what kind of things might happen as a result (fighting, new rules about sharing, complain to the principal about the teacher's elitist attitude, etc.). Compare these feelings to Peru's situation and tell students Peru's government has the same problem, and, as a result the army sometimes takes over the government.

3. Discuss the three regions found in Peru and the location of Peru in relation to the equator. Talk about the features of these regions: climate, geography, accessibility. How do these things affect population growth? (People tend to live in an accessible place and where the climate is not too extreme.) Where do most people in Peru live? (on a high plateau in the Andes) Where do you think might be the best place to live in Peru? (on the plateau where the climate is moderate) What kind of jobs are available in Peru? (fishing, farming, mining, oil production)

4. Assign **Handout 38** and have students use an atlas to complete their map of South America, using colored pencils to locate the various items and to complete a key for

their chosen symbols. Help students understand the various altitudes found in Peru. Tell them that some of the mines in Peru are up very high, and the miners chew the leaves of the coca plant (from which cocaine is derived) to help them adjust to the altitude and do such heavy work in such a thin atmosphere.

5. Use **Handout 39** with students after discussing the demographics of Peru. Be sure that students understand that the climates of the rainforest area and the high mountain regions are undesirable for most people. The coastal area is desert, so most of this area is also deemed unsuitable. Tell students that the government would like to see more settlement in some of the regions, particularly in the selva because mineral and oil deposits exist there. Without laborers to get them out, however, these deposits have no value.

   Have students create interesting help-wanted ads that might encourage people to come into these areas. Ads should be illustrated and contain a small map to show the city and surrounding area. Discuss how people adjust to their environment and remind students that people often live in a place they do not like because of employment opportunities.

6. Display the ads. Have students tell the class about their ads. Would the ads appeal to them enough to cause them to go there for work if they really needed money? Why or why not?

Enrichment/Extension

1. Research the Incas. Speculate about what changes there might have been if the Incas had discovered the wheel. Find out about their road system, the civil war between the half-brothers, and the causes of their defeat by the Spanish. Write and share with the class a report on your findings.

2. Work in groups to study Peru's economy today. What crops are raised? Where? For what purpose? What industries exist? Examine the underground economy of cocaine production. Why do Peruvians produce and sell drugs that all world governments frown upon? What does the U.S. government say about Peru's role in supplying illegal drugs to U.S. citizens?

Name________________
Date________________

Peru

Read about Peru and be ready for discussion questions.

Think rich, think poor, think Peru! This country, which is located in the Andes Mountains, has cities where many people live in a modern, well-educated, well-to-do society, but others, mostly Indians, walk those same streets barefoot or in sandals hoping to earn enough for food. There is a sharp division between the rich and the poor.

Spanish is the language used by government and business, but over half the population speak the Indian languages, Quechua and Aymara. Most of the country's economy is based on its farms, plantations, and mines. Most of these goods are exported to other countries. Peru also has an "underground" economy that is illegal, but probably brings in great amounts of money for those involved—the production of cocaine. Many rebellions have occurred in Peru because people disagree with the politics of their leaders. Often the army takes control and violently ends these class rebellions which have started because of the differences between the rich and the poor of Peru.

What kind of country is Peru? Peru is the third largest country in South America. Brazil and Argentina are larger. It is 496,225 square miles and lies between 5° and 18° south of the equator. It is bordered on the north by Ecuador and Colombia, in the south by Chile, in the east by Brazil and Bolivia, and the Pacific Ocean in the west. Peru has three distinct regions: the costa, along the Pacific Ocean; the sierra, in the Andean highlands; and the selva or la montana, which is found on the eastern part of the Andes and in the rain forest area of the Amazon River Basin.

The costa is mostly desert with only an inch or two of rain each year. The temperature tends to be mild—sixties and seventies—depending upon the season. Many earthquakes occur in this region. Buildings found in the capital at Lima are built in a special way to withstand these tremors. The river valleys are good places for farming by using irrigation. Peruvians produce a variety of crops. Fishing and oil production in offshore areas provide most of the jobs for those living along the Pacific Ocean. Lima and its port city, Callao, have about a quarter of Peru's total population and about 70 percent of its economic activity. Lima was built by the Spanish in 1535, and the Spanish style of architecture is still visible.

The Andes are often called the "Backbone of South America," and Peru has three of its ranges running through it. Some of its peaks are over 18,000 feet. The Quechua and Aymara Indians farm the mountain sides just as their ancestors did hundreds of years ago. In fact, they still use the irrigation system built by these ancestors. About half of Peru's population live in this part of Peru. This was the home of the Inca Empire.

The selva or la montana is a large tropical forest created by the heavy rains that are brought over the Andes by the northeastern trade winds. In the city of Iquitos 103 inches of rain falls yearly. The selva is found on the eastern slopes of the Andes (la Montana) and the land to the east (low selva). Although over half of Peru's territory lies in this regions, few people live there. Roads are not always passable, with many landslides adding to the problems. River travel is the best way to enter this region. Timber, oil, furniture, tropical fish, and plants are the chief exports of the Iquitos.

Name________________
Date________________

Most of Peru's economic problems are because it is underdeveloped and because of the great difference between the mountainous and the coastal regions. Most of the people in the mountains are poor, while the coastal regions are where development is taking place and thus, people there are better off. Peru has a population that is greater than it is able to feed so it must import a great deal of food. Transportation systems are poor—few good roads into the interior, and railroads that do not connect many of the cities—so efforts at uniting the nation and creating some kind of social equity at this time are not possible.

Name__________________
Date___________________

Peruvian Geography

On the map of South America, use a **brown** colored pencil to shade in the Andes Mountains. Using **green**, color the Amazon Basin. Use **blue** to draw the Amazon River, Orinoco River, Lake Titicaca, and Maranon River. Label the capital of each country in **red**. Color the oceans **blue**. Make a key using geometric shapes with color to indicate where the mineral and energy resources can be found on this continent—*e.g.*, silver might be a yellow circle on your map. Use an atlas to help you.

Name________________
Date________________

A Peruvian Ad

Compose a help-wanted classified ad with an illustration that would encourage people to come and work for you in one of the following positions. It should include the job requirements, pay, benefits, a picture that makes the region seem inviting, and a map showing the exact location of the city. Use color to make it more interesting.

Jobs

1. in the selva at a lumber mill
2. on the costa at an offshore oil rig
3. in the sierra, on a large cooperative farm

Classified Ad

Lesson 15
The United Kingdom—A Powerful Island Nation

Objectives

- To locate and identify physical features found in the United Kingdom
- To relate the United Kingdom's physical features to the human-environment interaction
- To chart important economic features found in the United Kingdom
- To examine an early time period in the United Kingdom and to compare it with modern times

Notes to the Teacher

In this lesson, geographic, historic, and economic aspects of the United Kingdom are examined and discussed. The two islands off the Western European coast are Great Britain and Ireland. Together they are called the British Isles. Great Britain is the larger island made up of England, Scotland, and Wales, while Northern Ireland is on the smaller Island (Ireland), composed of the six northern counties. Northern Ireland is sometimes referred to as Ulster. The United Kingdom has a long history with many different people contributing to its development as a nation. At one time the United Kingdom held a huge empire that spanned the circumference of the world.

The student reading gives a brief overview of the United Kingdom. Students use a blank outline map of the United Kingdom to locate and identify major geographic features. They examine the population of the large cities in this nation and its economic base by researching and completing a chart. The final activity is a historic one based on the feudal system found in the Middle Ages. Students research the system and its various elements and compare them with the modern United Kingdom.

Procedure

1. Have on hand some pictures or videos that show the various geographic features of the United Kingdom. Be sure to have a picture of the queen and other members of the royal family. Show them to students before they start to read. Find out what students already know about the United Kingdom. Do they know who is queen? Do they know her role in the United Kingdom?
2. Distribute **Handout 40** and have students read it. This can be done silently or in jigsaw cooperative groups with each group being assigned a piece of the handout. Then reform the groups with one member from each of the original groups becoming a new group. In this way, each newly-formed group has a member that was in each of the original groups. Students can share what they have learned as "experts" on the topic.
3. Display a wall map or transparency of the United Kingdom. Locate England, Wales, Scotland, and Northern Ireland. After discussing the political boundaries, locate the highland, lowlands, major cities, rivers, etc., with students.
4. Use **Handout 41** and have students complete the assignment with colored pencils. They are to locate and color code the various items listed on the handout. When this is completed, discuss with students how geographical features influence the way people use land. If it is too hilly, few people want to live or work there, yet mines are usually found in association with mountains. Someone has to work in the mines, therefore some people do move to these areas. Farmers prefer lowlands/plains areas to cultivate. Rivers and lakes are necessary to supply the water needed for agriculture, mining, and manufacturing. After the discussion, ask students to complete the assignment by writing a page or so to explain where certain economic activities might occur. As they use atlases, have students compare their ideas with maps showing agricultural and industrial production to verify if they were correct in their thinking.

5. Allow groups to complete **Handout 42**. Students need encyclopedias, almanacs, books on the United Kingdom, etc., to do this chart. Consider taking students to the media center.
6. Distribute **Handout 43**. Read the information about the feudal society and have students pull out the vocabulary words as they read. Let students talk about what they already know about this time period. Pictures of castles and knights are helpful.
7. Divide the class into groups of four to six students. (The list can be subdivided for the class if time constraints require the task to be shortened.) Let students complete the research on **Handout 43**.
8. Have students complete all the posters and projects as assigned. Be sure students have the materials they need. Consider having some tasks done as homework and others as classwork.
9. Allow student groups to make short presentations about their findings and to display their projects. Alert groups to be ready to respond to questions about their presentations.
10. Compare the government structure, economic system, military system, lifestyle, social strata, fashion, etc., with those of modern United Kingdom. What things have changed? What things have stayed the same?

Enrichment/Extension

1. Update **Handout 42** using the latest census. Compare and contrast any changes in the two charts.
2. Read about King Arthur or Robin Hood. Report, in oral or written form, on their lives and times and discuss them in relation to what you have learned in this lesson.
3. Research what is meant by an "English high tea." Organize a high tea for the whole class.
4. Explore the idea of constitutional monarchy. Find out which other countries have this type of government. Write an essay which speculates why the British people have kept the monarchy in spite of the bad publicity they have received in recent times. Share your essay with classmates.
5. Research Great Britain/United Kingdom and explain how it grew to be a huge empire and why that empire no longer exists. Gather pictures and information about some of the leaders in the countries who helped free themselves from the British Empire.

Suggested Responses, Handout 42: (1991 Census Data)

Largest cities	Population	Major crops	Major resources	Major products
London	*6,775,000*	*Barley*	*Coal*	*Automobiles*
Birmingham	*994,000*	*Oats*	*Oil*	*Machinery*
Glasgow	*687,000*	*Potatoes*	*Natural gas*	*Chemicals*
Leeds	*706,000*	*Sugar beets*	*Limestone*	*Ships*
Sheffield	*520,000*	*Vegetables*	*Iron*	*Electronics*
Liverpool	*474,000*	*Wheat*	*Salt*	*Textiles*
Manchester	*432,000*		*Clay*	*Banking*
Edinburgh	*438,000*		*Fish*	*Tourism, Ships*
Bristol	*392,000*	*Wheat*		*Ships, Flour Milling*
Cardiff	*388,000*	*Barley, Oats, Potatoes*	*Coal, Slate, Limestone*	*Coal Mining, Oil Refining*
Belfast	*374,000*	*Oats, Wheat, Barley, Potatoes*	*Grazing Land*	*Ships, Linens, Rope, and Twine*

Name________________
Date________________

The United Kingdom

Read the following article carefully.

At one time, it was said that the British Empire was so immense that the sun never set on it. There were British colonies all over the world. Through efforts toward independence, all of these colonies have left the British Empire to take charge of their own countries. What is the United Kingdom like today?

The United Kingdom today is made up of England, Wales, Scotland, and Northern Ireland. It is a changing country. Many years ago it relied heavily upon agriculture, mining, and the steel and shipbuilding industries. Today the farmland is being used for other things like highways and parks, and the business world has turned to the manufacture of electronic and computer equipment, automobiles, machinery, and chemicals.

Just as the world changes, so do countries. People are living longer because of medical advances. People are aware of their environment and the need to take better care of it. Shopping centers are moving into suburban areas and people leave the city stores to do their shopping at these centers. Trade is important because the United Kingdom has to import many of its natural resources to continue its manufacturing businesses, and the United Kingdom must export its manufactured goods and oil to other countries.

When the first census was conducted in the United Kingdom in 1801, most people lived in the country. London was the only really large city. A hundred years later, many people were moving into towns to take jobs in the manufacturing industry. They turned many of these small towns into large factory cities like Manchester, Birmingham, Glasgow, and Belfast.

The United Kingdom has a large population for such a small land area.

	Area	**Population**	**Capital**
United Kingdom	94,217 square miles	57.4 million	London
England	50,331 square miles	47.8 million	London
Scotland	30,410 square miles	5.1 million	Edinburgh
Wales	8,016 square miles	2.9 million	Cardiff
Northern Ireland	5,462 square miles	1.6 million	Belfast

*1991 census data

It is a crowded place. Almost 92 percent of the people of the United Kingdom live in cities and towns.

The United Kingdom lies off the northwestern coast of Europe—latitude 49° 57' N to 60° 50' N; longitude 1° 46' E to 8° 11' W. England is mostly rolling land. Moving northward toward Scotland the land rises to the highlands area that includes the highest point in the United Kingdom, Ben Nevis, at 4,406 feet above sea level. Lowlands lie in the center of Scotland. Wales is also a mountainous region. Northern Ireland is very rural and is mostly rolling hills suited for farming and cattle raising.

Name________________
Date________________

Climate varies in the United Kingdom. In the north and west of the United Kingdom the winters tend to be warmer and wetter than the south and the east. Summers in the south and east are usually hotter, drier, and sunnier than the rest of the United Kingdom. Warmth from the Gulf Stream ocean current keeps all of the United Kingdom warmer than other places at the same latitude. Fog is often a problem that causes travel to be delayed or causes many accidents. In the cities where pollution is heaviest, the fog often mixes with the smoke and dirt to create smog. The air quality becomes quite poor, and people may develop breathing problems as a result.

Over the years, many people have come to live in the United Kingdom or Great Britain as it is also called. Therefore, the island has a mixed population. Some of the first settlers were the Celts. They spoke languages known today as Cornic, Irish, Scots Gaelic, and Welsh. The Romans invaded the first century and stayed until early in the fifth century. Then came Germanic people—Angles, Saxons, Jutes. The English language spoken today was brought to the island by these Germanic people. The Vikings came and invaded the coastal regions, but, in 1066, William the Conqueror, a Norman, made the last invasion. The Normans were people from France who had descended from the Vikings.

The United Kingdom became a great sea power in the sixteenth century when it defeated the Spanish Armada in 1588. The United Kingdom's power grew until in the middle of the nineteenth century. The United Kingdom was the richest country in the world and had the largest empire. After World Wars I and II, British power slipped downward. Many of the colonies began seeking their independence, and, by 1980, most of the colonies were freed.

The government in the United Kingdom is a constitutional monarchy with a queen as head of state. The prime minister is the head of government. There is a legislative body called the Parliament with two houses: house of commons, elected by the people for a maximum term of five years, and the house of lords, hereditary peers (those whose titles come from parent to child) and life peers (those whose titles are just for them in their own lifetime). The government is formed within the house of commons where the political party with the majority of elected members decides who will serve as the prime minister. The prime minister selects a cabinet to help operate government business.

The United States, as a former colony of the British Empire, retains many similar characteristics of its former "mother country." The English language is spoken, some parts of the government system are like the British system, laws, cultural traditions, religions, and other parts of the old British Empire can be seen in the United States. The British Empire may have been reduced in size, but its influence can still be seen all over the world especially in the use of the English language as a primary language of government and commerce.

Name________________
Date________________

Mapping the United Kingdom

Using an atlas, colored pencils, and the map of the United Kingdom below, complete the following.

- Locate at least three rivers in each country. Color them **blue** and label them.
- Color the highland areas **brown**. Circle Ben Nevis.
- Locate at least eight major cities. Mark them with solid **black** dots.
- Locate London. Use a **star** to label it.
- Locate the lowland areas and color these areas **green**.
- Examine the features you have marked on your map. On a sheet of notebook paper, write an explanation about where the best areas to farm, raise livestock, build ships, and operate factories would be located. Give logical reasons for your choices based upon the geography and the population of the regions.

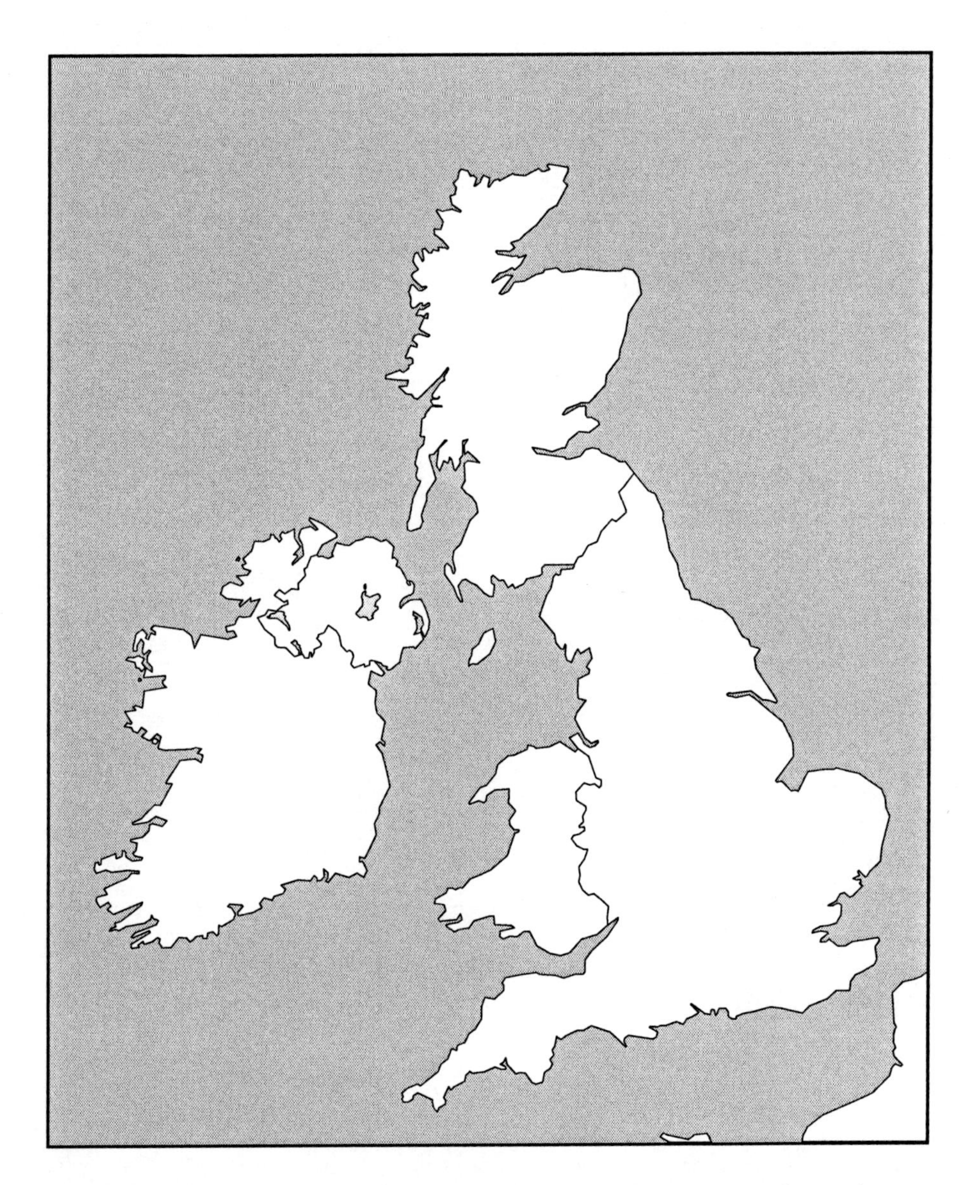

Name________________
Date________________

The United Kingdom

Complete the chart below by using encyclopedias, almanacs, or other reference materials about the United Kingdom. In addition to this information, make a list of the countries to which the United Kingdom exports its products and a list of countries where it buys its raw materials.

Largest cities	Population	Major crops	Major resources	Major products

Name__________________
Date__________________

When the United Kingdom Was a Feudal Society

About 1,000 A.D., during a time known as the Middle Ages, life in Western Europe was hard. There was no central government to protect or to support the people. Kings were not very strong, so they depended on the nobles for their food and their protection. These nobles were often knights who wore armor and rode horses to do battle with the enemies of their king. Nobles lived in castles. If a young noble wanted to become a knight, he would go to live in the castle of a great lord to receive the proper training.

Most of the people in the United Kingdom and other Western European countries were poor. These people, the peasants, did the farm work on the great manors. Manors were farming communities owned by the noble. This included the castle and a great deal of land surrounding the castle. The land was divided into strips for each of the peasants to work. Some of these people were serfs who were bound to the land and who could not leave. Some of these peasants were called freemen because they paid the noble for the right to farm the land. Together they supplied the food and tools needed by the knight and the king. There were no towns and therefore, no shops in those days. Everything that was needed was grown or produced on the manor. The knight protected the manor and the peasants as well as his king.

Your assignment is to work in cooperative groups of four to six students, in order to complete the following tasks:

- Find information about castles, knights, manors, feudal systems, and peasants.
- Make a poster that shows a floorplan/blueprint of a castle. Explain the importance of each feature—safety and/or comfort.
- Make a castle model using cardboard and other materials.
- Make a poster showing an English knight. Label the different parts of his armor.
- Make a poster showing the weapons that would be found inside the castle to protect it from enemies.
- Find out what kinds of fabrics were used and how they were produced. Make a drawing of the process and create posters of paper dolls illustrating the various fashions for kings, nobles/knights, ladies, and peasants in Great Britain around the year 1000 A.D.
- Prepare an oral report about the information your group gathered and present it to the whole class. Add other information that you found as you worked, such as foods they ate, the kinds of homes in which the peasants lived, their ideas about cleanliness, medicine, religion, etc.

Lesson 16
Scandinavian Peninsula—Norway, Sweden, and Their Neighbors

Objectives

- To compare the early Vikings with their peaceful, prosperous Norwegian and Swedish descendants
- To examine the ways that Norse and Swedes have overcome the limitations of their environment

Notes to the Teacher

Actually, the Scandinavian countries include not only Norway and Sweden but also Denmark, Finland, and Iceland. In the northern reaches of the peninsula, a fifth group, Laplanders, live near the sea in Norway, the rivers in Sweden and Finland. For centuries, Laplanders were nomads who learned to adapt to their harsh climate. Like the native Americans, they had no boundaries but lived with and shared the land of the reindeer they hunted and tamed. Today, in addition to breeding reindeer, they work in the mines, in hydroelectric plants, paper mills, and other industries which modern technology has developed in their areas.

This lesson, however, is focused on Norway and Sweden as the principal components of the large peninsula which juts southward from the Arctic circle. Their terrain is similar—mountains, rocky plateaus, great forests, jagged coastlines cut deeply by fjords (a long narrow arm of the sea bordered by steep cliffs), the interior laced with rivers and thousands of lakes. Sweden has more arable land on the southern tip than does Norway. Both countries make the best of their limited space and short season for growing crops. Their rich timberlands, always sources of building materials for homes and ships, have led to the development of great paper and pulp mills. Unlike others in the western world, every spring, they plant a hundred million saplings to replace what has been cut down. These are planted mostly by school children and university students. The northern mountains provide some of the richest iron ore in the world. Always among the world's great shipbuilders, Scandinavians have sent their ships of all kinds and sizes across the seas of the world. Farming, forestry, and fishing were the main sources of income which, in modern times, have been supplemented by more urban, industrial pursuits.

Students complete a map activity featuring aspects of Scandinavian geography. They then research either Norway or Sweden in order to create a poster demonstrating why the country is often labeled one of the most liveable and progressive nations in the world.

This lesson will take more than one day to complete. Text material on Scandinavia is an integral part of this lesson.

Procedure

1. Introduce the topic using material in the Notes to the Teacher.
2. Refer students to the map of Scandinavia in their textbook. Have them examine the map, especially the southern ends of Norway and Sweden. With a little imagination, they can see a resemblance to dragons. Just above the tip of the peninsula, the lakes in each appear to be eyes; the mountain range markings are like dragon's scales! In the eighth and ninth centuries when the Vikings sailed the northern seas in their dragon-headed ships, the Norse and the Danes terrorized northern Europe from Britain to France down to Spain. The Swedish Vikings turned eastward, sailed down the rivers of Russia to the Black and Caspian Seas, raiding the weak tribes and trading with the strong on the way. Some Vikings would stay to guard the ships while their companions followed the silk road to the Orient and back. Between raiding and trading, these Vikings, also known as Rus, established settlements, notably Kiev, became rulers, and eventually became assimilated with those they had conquered. Have students follow these travels on their map as the events are related.

3. Distribute **Handout 44**. Have available a world almanac for reference in answering questions 2 and 3. Allow students time to complete questions 1–4. Then distribute to each student one piece of black and one piece of white construction paper. Give them time to complete items 5 and 6. Ask several students to share their reflections on this activity.

   Suggested Responses:

   1. *Oslo, 60° N, 10° E; Stockholm, 59° N, 18° E; Trondheim, 63° N, 10° E; Malmo, 55° N, 13° E; Hammerfest, 70° N, 23° E; Copenhagen, 55° N, 12° E*
   2. *Temperatures in Oslo, Stockholm, and Copenhagen are relatively mild. In Oslo, recorded extremes are 93 and –21; Stockholm, 97 and –26; Copenhagen, 91 and –3.*
   3. *More typical temperatures in Oslo are average daily minimums of 20° in January and average daily highs of 73° in July; in Stockholm, average daily January minimums are 23° and average daily highs are 70° in July; and, in Copenhagen, average daily January minimums are 29° and average daily highs in July are 72°.*
   4. *Cities north of the Arctic Circle, such as Hammerfest, receive no sunlight at all on the day of the winter solstice, and the sun never sets on the day of the summer solstice. Days with either total sun or total darkness increase north of the Arctic Circle as one nears the North Pole, which has six months of sun and six months of darkness.*
   5. *Students' responses should recognize that extended periods of darkness or lack of light cause depression in winter.*
   6. *Students' responses might mention having the benefit of more time for outside activities, as well as the possibility of not getting enough sleep or having difficulty sleeping at night.*

4. Tell students that the descendants of the Vikings lost none of their ancestral courage or daring in creating and maintaining their most peaceable kingdoms at the top of our world. With the former U.S.S.R., a bristling armed camp on their back door step, and the U.S.A., equally armed, just across the Polar cap, Norway and Sweden have maintained their neutrality. Sweden has not been at war since 1814. Norway and Denmark were forcibly invaded by the Nazis in World War II in violation of their declared neutrality. During the five-year occupation, the Norse and Danes quietly joined the resistance and saved their own king and thousands of Jews. Even the children helped to save Norwegian gold by transporting it under their sleds to safe hiding places. Far from losing their courage and energy, they have channeled them into improving life for their own citizens. Presently, they have the highest living standards in the world.

5. Distribute **Handout 45** and give students time to complete the research and preparation of the posters. When the posters are turned in, take time to debrief the activity by asking students what they found most admirable—and least appealing—about their country.

Enrichment/Extension

1. Create a bulletin board display of the posters completed for **Handout 45**.
2. Research the kind of taxes and levels of taxation that support the lifestyle of the country you studied in **Handout 45**.

Name________________
Date________________

The Scandinavian Peninsula

1. Using the map in your textbook, note the latitude and longitude in which the following cities are located: Oslo, Stockholm, Trondheim, Malmo, Hammerfest, and Copenhagen.

2. What would the range of temperatures be in Oslo, Stockholm, and Copenhagen?

3. List the temperatures in winter and summer for each. Remember that the Gulf Stream influences southern ports.

4. What cities would be in the latitude where the sun shines all night in summer or it is dark all day in winter?

5. Now take the black piece of construction paper, stare at it with concentration for two minutes. After two minutes, write a paragraph in which you record your impressions of how you would feel if you had a twenty-four hour night for a good part of your school year.

6. Do the same thing with the white construction paper, but this time you are experiencing a twenty-four hour day during your whole summer vacation. What would be the advantages and disadvantages?

Name________________
Date________________

Scandinavia—Quest for Perfection

In 1984, a University of Pennsylvania professor concluded his ranking of 107 countries in terms of their general well-being. He used a total of 44 factors, including health, literacy, welfare programs, political stability, and status of women, among his standards. In his order of merit, Norway, Sweden, and Denmark earned three of the top five places.

Your task is to determine why one might consider a Scandinavian country a highly desirable place to live. Create a poster that demonstrates why either Norway or Sweden deserves to rank among the most livable places on earth. You may use pictures, charts, graphs, cartoons, or recent news items (or a combination of these) to make your point. Your poster should give attention to *any ten* of the following twelve categories:

a. Climate

b. Scenic beauty

c. Health care/life expectancy

d. Education

e. Career/job opportunities

f. Social programs

g. Status of women

h. Cultural opportunities

i. "Social Wage"

j. Ombudsman program

k. Standard of living

l. Approach to crime and punishment

Be sure to title your poster.

Resources you may find useful include the following:

Your textbook

Encyclopedia Britannica

World Almanac

Scandinavia (Chicago: Time-Life Books, 1987).

Issue of *National Geographic* with features on Scandinavia

Lesson 17
France—
The Linchpin of Western Europe

Objectives

- To infer how geography has shaped a country's development
- To identify France as an agricultural and industrial nation

Notes to the Teacher

France, one of the larger countries in Europe, has been called the linchpin of western Europe. Opening of the Chunnel has only increased the country's international importance. Recognizing that they are strategically important to the defense of the rest of the continent, the French have been able to pursue a somewhat independent foreign policy. However, the topography of France has promoted the unity of the French people. Mountains and coast lines provided most of the country's natural boundaries. It was easy to build canals, highways, and railroads on the low, level land. The people of France had no great difficulty in traveling from one part of the country to another; thus, they grew to know one another and they learned from one another. Sharing the same language and the same ideals and customs, differences in the people began to disappear, and Paris became the hub leading from one part of the country to another.

France is the only nation in western Europe that is agriculturally self-sufficient. More French people are engaged in agriculture than in any other occupation. Many of those who do not actually work on farms handle or distribute the products raised by farmers. Most of the land of France is devoted to crops and meadows. The farms are small, averaging about twenty acres, and most of the farmers own their own land on which they work. Handed down from father to son, many farms may remain in the same family for many generations.

France has a great variety of farm products because it has a variety of lands and climates. Sugar beets and flax are suited to the cool plains of the north. Wheat thrives on the rich soil of the Paris Basin. Oats and rye do well in the cool, wet highlands of the Central Plateau. Summers in the Garonne Valley of western France are long enough for corn to ripen. Grapes grow on the sunny slopes of river valleys. Dairy cows are raised on farms where grass grows well. The dry, chalky lands of the Champagne region are excellent for grazing sheep and both sheep and goats can live on the scanty vegetation of the Central Plateau.

French industry combines heavy manufacturing with high quality luxury products. France's iron and coal deposits supply three major steel making centers. The French, known for their artistry and skilled craftsmanship, produce luxury items with pleasing colors and beautiful designs. Noteworthy exports include their fine glassware, porcelain, furniture, textiles, and leather products.

Paris, located in the heart of a fertile farming country, near the junction of the Seine and Marne rivers, is the country's primary city. An island in the Seine provided natural protection from enemies of its ancient inhabitants and serves as the center of modern-day Paris. The city's strategic location has helped establish the city as the center of the country's trade, transportation, and culture.

Students begin by drawing as many inferences as possible about this as-yet-unidentified country from a few basic geographic facts. They then test their inferences against the realities of modern-day France and use discussion and two additional handouts to develop their understanding of how France's geography largely determines its importance as the linchpin of western Europe.

Procedure

1. Divide the class into small groups and distribute **Handout 46**. *Do not* reveal the name of the country at this point. Give students five to eight minutes to make as many inferences as possible about the country's probable development.

2. Compile a class list of inferences. Share information from the Notes to the Teacher and reveal the country so that students can check the accuracy of their inferences.

3. Have students use their textbook map or have copies of a map of Europe available for each student. Tell students that they are going to use the map to find some physical characteristics that make this region prosperous in agriculture and industry.

4. Ask students to locate the Rhine, Seine, Saône, Loire, and Rhone Rivers and ask, "How have these rivers led to the development of trade and the exchange of ideas?" (These rivers make it easy to transport goods between regions in France and other countries. They are important commercial and industrial routes for Europe.)

5. Have students use their maps and trace the course of each of these rivers (Rhine, Seine, Saône, Loire, and Rhone), noting the areas through which each flows and the body of water into which each empties. Tell students that the French have built many canals that connect these rivers.

6. Distribute **Handout 47** and have students explain the landforms listed. Using a dictionary or the glossary in their text, have students complete the handout. Use descriptions and diagrams of each landform on the board.

   Suggested Responses:

   1. An area of fairly level high ground

   2. Area of level land along the seashore

   3. Elevated or mountainous land

   4. Usually rounded natural elevation of land, not as high as a mountain

   5. A mass of land higher than a hill

   A. Mountains

   B. Hills

   C. Plains

   D. Plateau

7. Discuss the following points:

   a. What features made Paris the primary city of France?

   Paris was located in the heart of the country near highways, railroads, canals, and navigable rivers.

   b. Why do so many French industries specialize in small, high-quality products?

   French ships cannot sail in the rivers for long distances; goods have to be sent by railroad to seaports of other European countries before they can be shipped overseas. Moving large goods is too expensive; the French specialize in goods that are high in value but low in weight.

   c. What products might France export to the United States?

   Clothing, glassware, porcelain, furniture, textiles, cheese, wine, jewelry, and soap (partial list—others can be added)

8. Conclude the lesson, by assigning **Handout 48** to give students a chance to demonstrate their grasp of key geographic terms covered in the lesson. Provide class time to share students' rewrites.

Enrichment/Extension

1. Play French language tapes and see how many words you recognize. Develop a booklet of French words with the English equivalents.

2. Visit the library to find books about France. Do a book report on one.

3. Visit travel agencies to collect posters and brochures on France. Make a display for your classroom.

4. Look for pictures and articles to illustrate the following about France:

   a. Physical features and climate

   b. Agricultural products

   c. Resources

   d. Industries

   e. Important cities

   Make a bulletin board of what you find.

Name_______________
Date_______________

Geography's Impact on Development

At this point in your study of world cultures and geography, you have considerable knowledge of how the two are related. Listed below are several key factors of the geography of Country X. In your small group, brainstorm a list of ways that these geographic factors have likely affected the country's development. You should be able to make many inferences about the country's strategic importance, economy, society, culture, recreation, and tourist possibilities.

Country X

Area	210,026 square miles, about 4/5 the size of Texas
Neighbors	borders six other countries
Highest point	15,771 feet
Ports	5 major ports
Canals and rivers	7957 kilometers of navigable canals and rivers
Resources	bauxite, iron, coal, crude oil, forests
Topography	wide plain covering more than half of the country
Climates	oceanic, continental, and Mediterranean

Name__________________
Date___________________

What Do You Know?

France has many farm products because it has a variety of lands. Can you explain the landforms listed? You may want to do some research before you try to describe them. Write your descriptions in the space provided and label the diagram with the landform that it represents. Note: Highlands do not appear in the diagram.

1. Plateau(s)

2. Plains

3. Highlands

4. Hills

5. Mountains

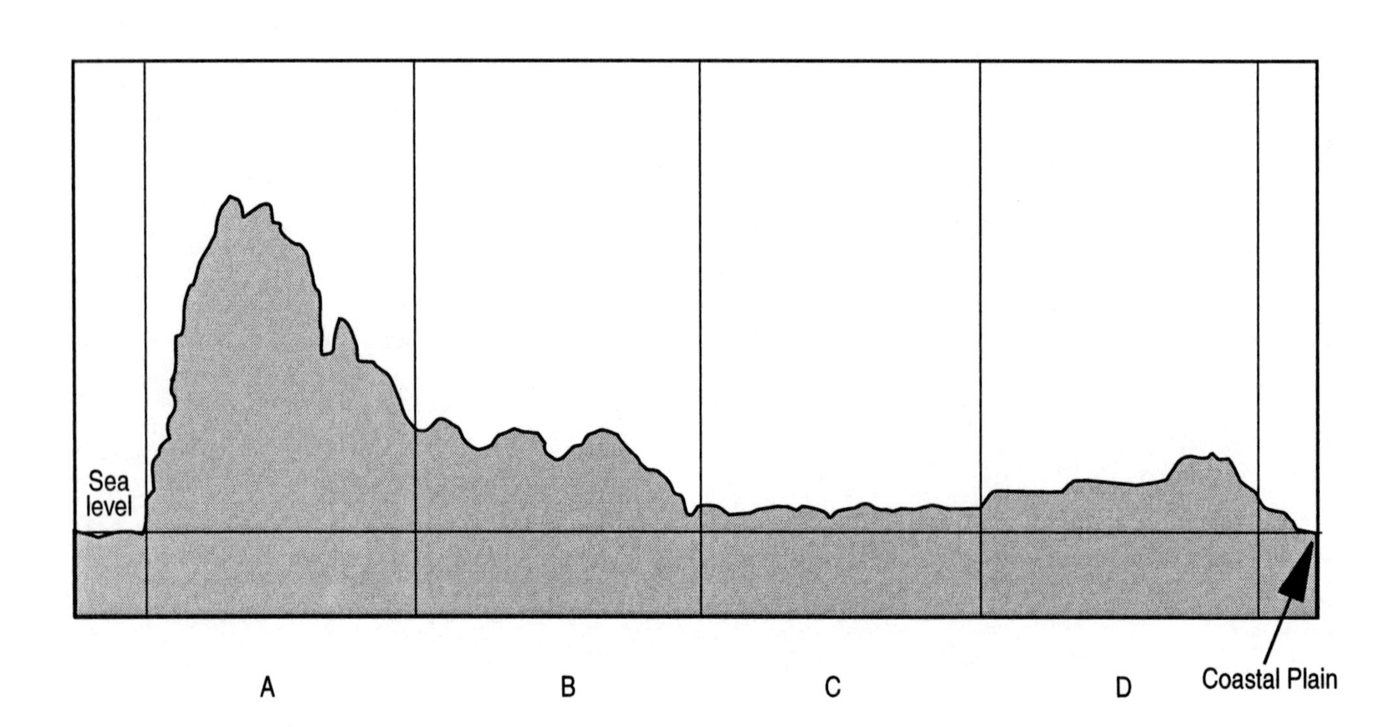

Name________________
Date________________

Building Your Geographical Vocabulary

In the space provided, rewrite the sentences to show that you understand the definition of the italicized words. Use your dictionary or the glossary in your textbook.

1. The French have built many *canals* to connect their *navigable* rivers.

2. In France, as in most countries, there is a *primary* city.

3. Some regions in France have a *dense* population.

4. Many products are *exported* from France.

5. France has two long *coastlines*.

6. Rivers were important to the development of French *industries*.

7. French industries *specialize* in small, high-quality products.

Lesson 18
Iberian Peninsula—Spain and Portugal

Objectives

- To compare post-empire Portugal with post-empire Spain
- To investigate how these two countries with similar problems have come to grips with them

Notes to the Teacher

The countries of Portugal and Spain are linked not only by geography, but by history as well. Both countries were at one time great empires. In fact, parts of what are now the United States were once part of Spain's empire. Both countries continue to play vital roles in European and global affairs.

Students complete a reading and map activity. After discussion, they participate in a culminating activity.

Procedure

1. Bring in an empty wine bottle. Ask the class where the wine came from. (answer determined by reading the label) Uncork the bottle and ask where the cork came from. (It is doubtful if anyone will know. If someone does know, ask how they know.) Inform the class that regardless of the country of origin of the wine, the cork for the bottle, undoubtedly came from Portugal. (Alternate introduction would be to use the cork bulletin board if the classroom has one.)

2. Distribute **Handout 49** and have students complete it as directed. Some possible discussion questions are as follows. Little Portugal once claimed as part of its empire one of the largest countries in the world. Name that country. *(Brazil)* On which continent is the former colony located? *(South America)* What is the capital of Portugal? *(Lisbon)* What is fado? *(Portuguese folk music)* Name a state of the United States that was once part of the empire of Spain. *(California, Florida, etc.)* What is the capital of Spain? *(Madrid)* Who is the king of Spain? *(Juan Carlos)*

3. Debrief the map portion of the handout by locating the significant places, rivers, and bodies of water on a wall map or overhead transparency map.

4. Distribute **Handout 50**. Explain that these topics on which students are to find as much information as possible will be the basis for a game of Tic-Tac-Toe. Have a student draw a nine-section game board on an old sheet with a broad-tipped black magic marker. Have the student mark the nine sections as: (1) Odds and Ends, (2) History, (3) Colonies, (4) Romans in Spain and Portugal, (5) Moors in Spain and Portugal, (6) Places of Interest, (7) Products and Industries, (8) Artists, and (9) Writers. Have another student make sets of Xs and Os.

5. Divide the class into six groups. Assign each group one or two categories on which to compose a minimum of twenty questions with answers. Have groups print each question on one side of an index card and the answer on the back. (Depending on class size and ability, you may wish to have nine groups of two or three students each who have one category to research.) Preparation time will depend on class ability. On the assigned day, appoint an emcee to ask questions and use a timer with a push-button bell. Ask for two volunteer contestants. To give all students a chance to play, set a five or ten second limit for each answer and a three or five minute game period. Students who have prepared the questions may decide whether a disputed answer is acceptable. Instead of money prizes, students may earn bonus points. The teacher may decide the number of points needed to be excused from taking a unit test or to be declared champion.

Enrichment/Extension

1. Prepare an Iberian cookbook. Bring in food made from these recipes and share it with classmates if permissible.

2. Prepare an oral report (5 minutes) on the tiny independent state of Andorra, the Basque people of Northern Spain, or the British Crown Colony of Gibraltar.

Name________________
Date________________

The Countries of the Iberian Peninsula

Read the selection below and be prepared to discuss it. Then, complete the map activity as directed.

Part A.

Portugal, which occupies only one-sixth of the Iberian peninsula, once ruled an empire twenty-three times larger than her own landmass. This empire, of course, included Brazil in South America, colonies in Africa, and islands in the China Sea. The break-up of the empire in the twentieth century impoverished Portugal to the point where it was the poorest country in Western Europe. Today, however, the Portuguese people, with their determination and individual pride, are making a comeback. These descendants of fearless navigators and explorers who plotted ship routes around the globe are rebuilding their economy along with their national pride. Their capital city of Lisbon, once demolished by an earthquake, is now a thoroughly modern complex of high-rise office buildings with orange tiled roofs surrounded by its old city which has preserved its traditional local color.

Strolling along the old quarter, one can see five hundred years of Moorish influence in the archways and latticed windows, smell the fragrance of almond and lemon trees, or hear the lilt of fado music. As one of its greatest singers, Gina Guerra, describes it: "Fado deals essentially with love and therefore with sorrow. The two are inseparable, and it is merely that bittersweet quality of life, not hopelessness, that Fado portrays."[1]

Portugal now has the world's largest cork industry as well as a corner on the market for its internationally prized Port wine, a large sardine fishery, and a 2.4 billion dollar tourist trade. In addition, billions of dollars are sent home by Portuguese working abroad. In the North there is a shortage of jobs which leads people to work in the industrial countries of northern Europe temporarily. Paris alone has half a million resident Portuguese workers. It was, however, the small farmers and grape growers who saved the elections of 1975 from a Communist victory. The latter needed the north to carry the election but made one fatal error in their campaign. The Communists promised these small farmers (who had just acquired their own land from taking over the property of the great landowners) to abolish all private ownership.

All problems have not been solved as yet, but with the Portuguese spirit of hard work and determination, the country may make a success of its parliamentary democracy.

The sister country of Spain, occupying five-sixths of the Iberian peninsula, has a similar story of a lost empire which once included all of the South American countries except Brazil, Mexico, the southwestern United States, Cuba, Puerto Rico, Central America, colonies in Africa and the Philippines. It has also experienced a decline, a bloody Communist revolution, a longer dictatorship (France ruled for forty years) and now an era of prosperity in a constitutional monarchy.

Present writers are not so much concerned with the "rain in Spain" but with the change in Spain. Madrid, once so isolated from the rest of the country on its high, dry plateau that we wonder why it was chosen for the capital, has, in this century, become an active member of this changing world. Sleek, modern trains and jets have made it accessible to its European neighbors, the western world, and its own citizens. Not only millions of tourists have poured into Madrid and all of Spain, but also much of American and European fashions, entertainment, and social changes are reflected in a

[1]William Graves, "After an Empire . . . Portugal," *National Geographic*, (Dec. 1980): 811.

Name________________
Date________________

growing, prosperous, middle class, in more and more women in the business world, and in improved organization of the workers. Traditionally, life begins after a 10:00 P.M. dinner, as Madrileños set out for a night on the town. As they stroll about the city, they may see anything from a high-wire performer on a motorcycle 300 feet above the Plaza D'Espagne or even a way-out fashion show featuring styles based on the latest science fiction movies or they may spend their remaining hours at a discoteca of their choice. Such is "Mod"-drid!

The population of Madrid has grown from 1.6 million in 1950 to three million in 1991. Since the city cannot provide living space for all, suburbs have multiplied on the outskirts. Migrants from the poor south of Spain often live in substandard shelters but move as soon as they can to modest but adequate housing. The more affluent middle class has established its own suburbs. Even the royal family lives on the outskirts of Madrid at Moncloa Palace. They use the huge royal palace in the capital only for state occasions and open it to the public as a source of tourist income. Not wishing to destroy their heritage, Madrileños have preserved and restored their fine old buildings, cleaned up their parks, planted new trees, encouraged city planning that will provide some order for burgeoning suburbs, and provided better housing for workers.

In northeastern Spain, the flourishing seaport of Barcelona, center of Catalan culture, is also growing in population and prosperity. Once of Spain's greatest industrial centers, Barcelona, has extensive textile and shoe factories, automobile plants which turn out a million cars a year, and a shipbuilding business which makes Spain the sixth largest ship maker in the world. Andalusia in the South claims historic Seville, Grenada, and Cordoba which annually draw a large portion of the country's 38,000,000 tourists—Spain's new source of gold. Toledo, an ancient walled city, home of El Greco, specializes in the steel of famous Moorish swords, handcrafted damasquin gold and black jewelry, and tourism. As in Portugal, choice wines are produced in LaMancha and other areas. Jerez, home of the world's best sherry wine, does a thriving export business.

Nothing is perfect, however, least of all post-revolutionary Spain. Memories and scars of the Civil War (from 500,000 to 1,000,000 people were killed) are still there, especially among the older generation, and are quite visible in the number of handicapped people who run small businesses or work at compatible jobs. José Maria Aznar Lopez, a member of the Popular Party (PP), became prime minister in 1996. The popular King Juan Carlos is a man sensitive to all the conflicting issues of his society. He has been a stabilizing and unifying force in setting Spain on a true course into the modern world.

Name________________
Date________________

Part B.

Locate and label the cities mentioned in the reading. Locate and label the principal rivers and surrounding bodies of water. Use information from your textbook to mark the products of the different areas of the two countries.

Spain and Portugal

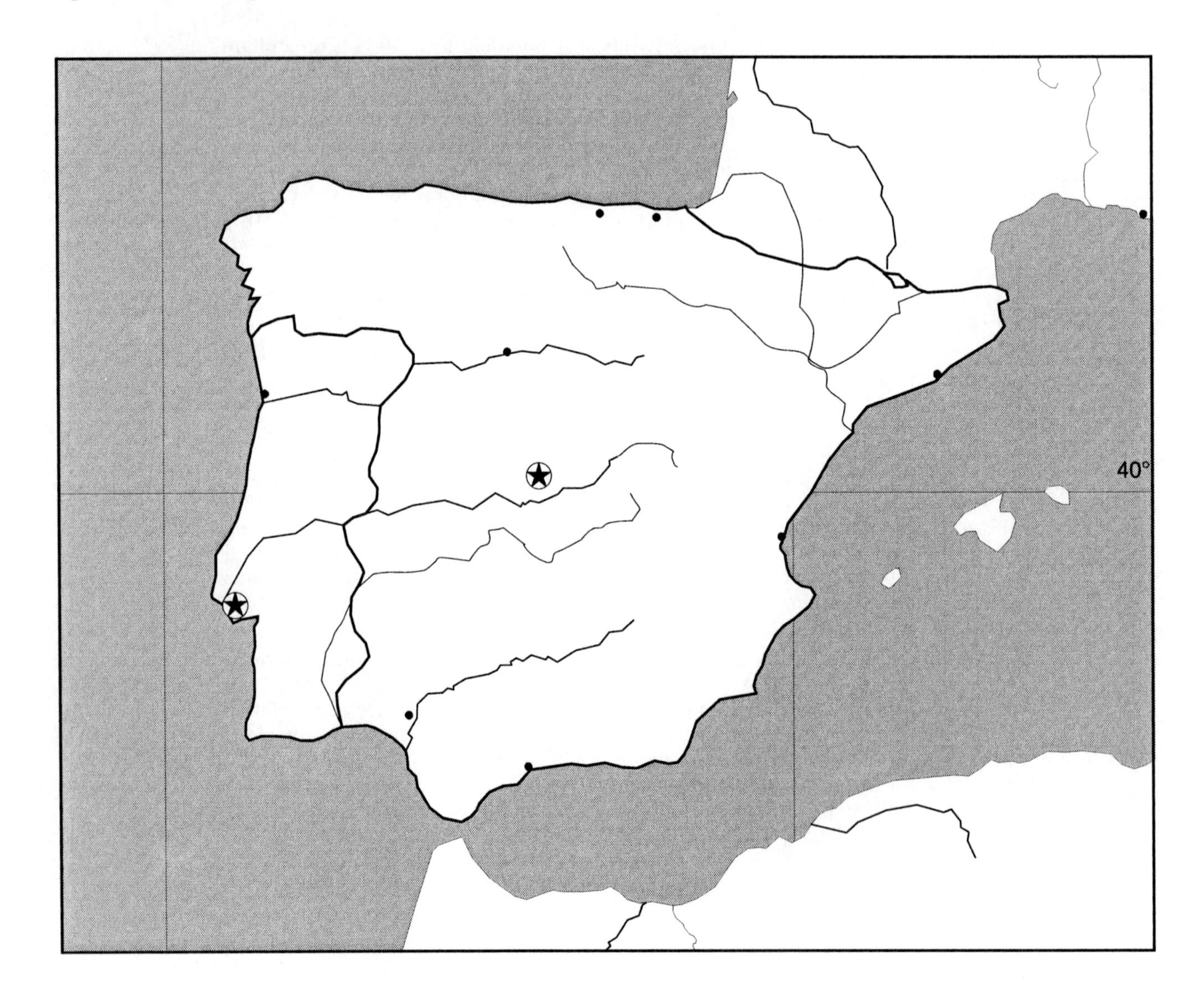

Name_______________
Date_______________

Iberian Tic-Tac-Toe

Suggested topics for game categories

1. **Odds and Ends** Flamenco dancing; bull fighting—difference between Spanish and Portuguese styles; January 6—Magi give gifts; Easter in Seville; Fado singing in Portugal; Cordoba's Kissing Street; Alcazar—model for Disney's castle; statue of a small bear in the center of Madrid's Puerta del Sol—its significance; El Rastro—enormous Sunday morning flea market (yard sale); cafes—part they play in city and village life

2. **History** Ferdinand of Aragon, Prince Henry the Navigator, Isabella of Castile, Don Juan of Austria—half-brother to King Philip II, General Franco, General Salasar, Inquisition, Magellan, Spanish Armada, 1936 Revolution, Vasco de Gama

3. **Colonies** Brazil, all of South America, Central America, Mexico, Southwestern United States, Cuba, Puerto Rico, African colonies

4. **Romans in Spain and Portugal** Province of Hispania; Latin language; Roman citizenship; engineering—bridges, aqueducts (Segovia) roads; amphitheaters; four later Roman emperors born in Spain (Hadrian, Marcus Aurelius, Trajan, Theosius); Roman writers (Seneca, Martial, Quintillian)

5. **Moors in Spain and Portugal** brought back Greek culture from Alexandria; historians, philosophers; grammarians; Arabic numbers; trigonometry; botany; medicine—pharmacology; architecture

6. **Places of Interest** Madeira and Azores Islands, Mallorca, University of Coimbra, Lisbon, Belem, Oporto, Fatima, Braga, Prado, Escorial, Valley of the Fallen, Royal Palace in Madrid, Plaza Mayor, Puerta del Sol, Toledo and Avila—two walled cities, Alhambra, Gold coast—Costa del Sol, St. James of Compostella

7. **Products and Industries** Wines, wool, oranges, olive oil, figs, automobiles, hydro-electric power, fishing, food processing, cork, cement, mercury, zinc, copper, lead, manganese, gold, silver, wolframite, pyrite, sulphur, resin, coal, iron, shipbuilding

8. **Artists** Picasso, Miró, Dali, Murillo, El Greco, Velaszuez, Goya, Ribero, Zubaran

9. **Writers** Calderon, Cervantes, Lope des Vegas, Teresa of Avila, John of the Cross, Garcia Lorcas

Lesson 19
Italy—A Heritage of Splendor

Objectives

- To examine the geography of Italy
- To experience the wonder and excitement of major cities of the Italian peninsula

Notes to the Teacher

Towering Alps to the North, a stiff backbone of the Apennines running down its length, a circle of dormant and active volcanoes curling around its toe, good harbors on both east and west coasts, Italy literally has its boot on the Mediterranean world. Once the heart of an empire stretching north to Britain, east to the Danube, west to Spain, and south covering the northern coast of Africa, making the Mediterranean a Roman lake, Rome ruled the known world. When the political empire collapsed under barbarian invasions and internal corruption, Christianity took over and made Rome the center of western medieval Europe unified by religion and the Latin language. When forces of unrest and change swept across the west with the return of the Crusaders, the Italian Renaissance, born in Florence, was to bring about the greatest cultural flowering of civilization since the golden age of Rome and Athens. Modern Rome, seat of the present government, mecca of tourists from every quarter of the globe, is a post of World War II miracle of revival which extends throughout the peninsula.

Students complete a map of Italy and consider ways the country's geography has influenced its development. To conclude the lesson, small groups prepare a brief commercial highlighting attractions of one of Italy's modern cities.

Procedure

1. Have students read the material provided by their textbooks on Italy. Stress the splendor of Italy's past by sharing material in the Notes to the Teacher that highlight the four major periods of the country's rich history.
2. Assign **Handout 51** either for homework or as an in-class activity to familiarize students with essentials of Italian geography. Have students use textbooks as a resource for completing this map.
3. Point out that geography has often played a critical role in Italy's development. Conduct a discussion that stresses the role of geography in instances such as these:
   a. Why was the city of Venice built on small islands connected by canals? *(Protection from barbarian invaders)*
   b. Why is the best farming in the northern part of the country? *(Broad plains of the Po Valley have ample opportunity for irrigation.)*
   c. Why did the eruption of Vesuvius in 79 A.D. give us our best indications of everyday life in ancient Italy? *(Fast-moving lava flows and the spewing volcanic ash buried low-lying Pompeii and Herculaneum without warning.)*
   d. Why has the island of Sicily been important for both military and trading purposes? *(It is an easy steppingstone to Africa.)*
   e. Why did Hannibal's invasion from the north come as a surprise in the Punic Wars? *(Italians believed the Alps offered sufficient protection.)*
   f. Why did the Italians have a monopoly of European trade with the Far East before the Age of Exploration? *(Strategic location almost required traders coming by land from the Far East to stop in Italian cities, allowing Italians to become middlemen for goods destined for points farther west.)*
   g. Why didn't Italy develop city-states as did Greece? *(The Apennines did not cut off trade, travel, and communication in the same way that higher mountains in Greece did.)*
   h. Why did the Italian peninsula become an agricultural rather than a seafaring and trading nation? *(Broad river valleys suitable for farming and few good harbors)*
4. Have students refer to their completed maps as differences among major cities of Italy are pointed out using the following information:

a. To the north, Milan is a flourishing industrial center. Not only is Milan the second largest city in Italy, leading the country in production of textiles, chemicals, metal products, printing, publishing, and banking, but it can claim the most famous opera house in the world, La Scala, and among their great art treasures, Leonardo da Vinci's *The Last Supper*.

b. "Bride of the Adriatic" is Venice, built by refugees from invading barbarians in the fifth century A.D., Venice dominated the trade routes to the East during the Middle Ages and Renaissance. Consequently, she grew wealthy from trade of silk, stains, brocades, gold, and spices. Venetian merchants built palaces and churches and employed the best artists to decorate them. Ironically, the sea, source of Venetian wealth and fame, is today's greatest threat to its existence. Motorboat traffic, industrial pollution from the mainland, and changing tides are undermining its foundations.

c. Moving down the peninsula. we come to Florence, the flower of the Renaissance. This small city on the Arno River, at the height of the Middle Ages, produced such poets as Dante and Petrarch, and, in the Renaissance, such a roster of artists, goldsmiths, sculptors, princes, and bankers which has never been equaled. The medieval Ponte Vecchio is crammed with modern shops, but when we look up, we can see Brunellischi's dome of the Cathedral, or walk a few blocks closer to view Ghiberti's Baptistry doors. In the Academic museum, Michelangelo's great statue of David is the central attraction.

d. Mistress of the known world from classical times to nearly the end of the fifth century A.D., Rome suffered its dark ages until it became the center of Christendom at the peak of the Middle Ages. Renaissance master sculptor, architect, and painter Michelangelo was summoned from Florence by the Pope to decorate the Sistine chapel, to sculpt the great statue of Moses, and design the new Basilica of St. Peter's. Today, Renaissance palaces, medieval churches, the classical Colosseum and Forum Romanum are interspersed with modern hotels, shopping malls, and traffic. Busy and crowded as it is, Rome has the independent state of Vatican City within its boundaries and territory belonging to the Knights of Malta, both completely independent.

e. Jewel of the South, Naples hugs its mountain and overlooks its deep blue bay. We can see smoking Vesuvius and the dormant Solfotaro on the horizon, the isle of Capri across the harbor. More than its noisy, narrow streets with incredible traffic jams, Naples has beautiful churches, palaces, and museums filled with art which reflect its variegated history. Its great opera house of San Carlo is second only to Milan's La Scala and its aquarium is one of the best in the modern world.

5. Divide the class into small groups with each group assigned one of the following cities: Milan, Venice, Florence, Genoa, Rome, or Naples. Tell students to pretend that their assigned city is their own community. If any students' families come from one of the cities or surrounding areas of Italy, be sure that they get into that group.

6. Distribute **Handout 52**. Remind students that they will have to look up information in other sources beyond their own texts such as encyclopedias, world almanac or yearbook, and other geography books for population statistics. Interviewing other people who came from that area or students who have traveled there would be helpful. The travel section of their Sunday newspaper is another source.

7. Conclude the lesson by having class presentations on the cities by either volunteers to the whole class, or by each individual in a small group (five to six students) as class time allows.

Enrichment/Extension

1. Research the city of Pompeii and prepare a five-minute oral report to share with classmates.

2. Research a famous Italian of any age (Julius Caesar to Federico Fellini). Write a biography of your chosen person and share it with the class.

Name__________________
Date___________________

Italy

On the map below, locate and label the following places:

Seas Adriatic, Ionian, Mediterranean, Tyrrhenian

Islands Corsica, Sardinia, Sicily

Mountains Alps, Apennines, Mt. Vesuvius

Rivers Po, Arno, Tiber

Cities Venice, Milan, Genoa, Florence, Rome, Naples, Palermo (Sicily)

Name__________________
Date__________________

Promoting Your City

Assume that you head the local Chamber of Commerce in one of the following major cities: Milan, Venice, Florence, Genoa, Rome, or Naples. It is your task to prepare the script for a three-minute commercial to advertise the merits of your city to potential newcomers and businesses. Make your city as attractive as possible to customers. Use these questions as guidelines.

1. Where is your city? Give the location in relation to another well-known place.

2. When was your city founded? Why did the first people choose that location? What geographic advantages did it have?

3. Has the location of your city had any effect on its growth and development? Explain.

4. Have there been any great changes in population? Why or why not?

5. What is the elevation of your city above sea level?

6. What are the average temperatures in January and July? The annual rainfall?

7. What are the most important industries? job opportunities? What makes your city different than other Italian cities?

8. What kind of schools does your city have? Any great universities? Museums? Ancient ruins? Traditions?

9. What opportunities for fun does your city offer? Describe them.

10. Italians love to have festivals or celebrations. What are some of the traditional fiestas that your city celebrates? Describe them.

For extra credit, you may also choose to prepare a travel poster with pictures of attractions your city has to offer.

Lesson 20
Greece—Birthplace of Democracy

Objectives

- To note the geographic factors which helped the ancient Greeks create their unique civilization
- To explore modern Greece

Notes to the Teacher

This small mountainous country in the Mediterranean is considered by some people to be the cradle of Western civilization. Few other countries of comparable size and location have left such a deep impression on the rest of the world. Limited farmland, coupled with a lack of mineral resources, almost required Greece to become a trading nation. Blessed with many harbors and hundreds of small islands nearby, early Greeks established trading partnerships with Cyprus and Syria to the east, Egypt and Libya to the south, and Italy and even England to the north and west. Widespread contacts with the outside world brought new ideas, among them the Phoenician alphabet, the Lydian system of coinage, the Babylonian concepts of astronomy, the iron working process of the Hittites, and the knowledge of medicine and geometry of the Egyptians. Creative Greek thinkers adapted ideas such as these to reach new insights in philosophy, architecture, art, literature, theater, science, mathematics, and medicine, particularly in the Classical Age of fifth century B.C. Perhaps the Athenians' greatest gift to us was their early concepts of democracy and the dignity of the individual, ideas we have, in turn, adapted and refined in a manner far beyond their original thinking.

Geography influenced Greece's development in other ways as well. The mountainous terrain kept Greece divided into smaller regions than promoting unification. Independence became a way of life and freedom, their prized possession. However, this fierce sense of independence brought problems. Ancient Greek city-states warred among themselves and united only to fight off common enemies. Frequently, only the Olympic games brought a truce over the entire peninsula.

Students work in small groups to create imaginary tours of modern-day Greece designed for special interest groups. In addition to plans for an eight-day tour of Greece and nearby islands, students draw or trace a map of Greece with highlights of their tour pinpointed, create a travel brochure, and use an almanac to compile a fact sheet of important information about contemporary Greece for their tourists.

Procedure

1. Refer students to a wall map or map in their textbook and have them locate Greece and the surrounding islands. As students note the many islands, large bodies of water surrounding Greece, and rugged coastline, ask them how these features help to explain why the Greeks became and still are a great maritime people. Ask, too, how the mountains and lack of resources have affected Greek development. Share the material in Notes to the Teacher.

2. Ask students to speculate why nearly 340 of the 500 Greek islands are uninhabited. (*An insufficient water supply and lack of arable land make most of the islands incapable of supporting human and animal life.*) Point out that the largest island, Crete, had an advanced civilization (the Minoan), while the people on the mainland were still primitives. The great palace of the king at Knossos even had indoor plumbing!

3. Have students locate Olympia, site of the games begun in 776 B.C. and then Mt. Olympus, the highest peak in Greece. Ancient Greeks considered this mountain in the north the home of the gods.

4. Distribute **Handout 53** and divide the class into small groups to prepare their tours of Greece and some of its nearby islands. Have each group plan an eight-day trip through Greece with special attention to the interests of their clients. When students have completed their final itineraries, have each group leader present the project

and tour plans to the whole class. Have students select the best tour and the best overall project and explain their reasons for those selections.

Suggested Responses, Part A (Based on 1995 data):

Government—*presidential parliamentary republic*

Population—*10,565,000*

Area—*50,944 square miles*

Religion—*Greek Orthodox*

Economy—*livestock (sheep and goats), agriculture (grains, cotton, tobacco, olives, and fruits), tourism, mining (bauxite, oil, lignite), textiles, and fishing*

Major cities—*Athens, Thessoloniki*

Climate—*hot, dry summers; cool, rainy winters*

Literacy rate—*men, 96 percent; women 89 percent*

Enrichment/Extension

1. Prepare a poster showing different kinds of regional costumes.
2. Research a visit to Mt. Athos, the monks' territories. Remember, no women allowed!
3. Find modern ads or political cartoons which use figures from Greek mythology, *e.g.*, Ajax for cleaning powder, to help sell the product. Prepare a poster with a minimum of six samples and a caption for each explaining the appropriateness of the myth to the ad.
4. Write an essay explaining why the two statements from the Delphi shrine are appropriate to Greek culture: "Know Thyself" and "Nothing in Excess." Examine some of the architecture to support your points.

Name________________
Date________________

Custom Tours of Greece

Your group operates a tourist agency, Homeric Adventures, Inc., that organizes tours of Greece customized to the interests of your clients. As organizers for the company, you have the task of planning an eight-day trip to Greece and one or two of its islands for a group that includes an architect, a mythology enthusiast, a theater director, a farmer, a fisherman, and a history buff. All of your group tours try to give participants the flavor of the country and a chance to learn the beauty and diversity of its scenery, a bit of the country's rich history, and its religion. Tours try to convey a sense of the people and their occupations, the food and dress of the people, their recreation, their homes, and specialty items available in their shops. To the extent possible, tour directors try to include places and people who will give travelers the most information and inspiration to do their own jobs better.

When completed, your project on Greece will have four parts:

Part A. A handout of the basic data about Greece and its people today

Part B. A map of the country showing places your tour will visit

Part C. An eight-day tour itinerary

Part D. A travel brochure, preferably with pictures, that entice people to sign up for your tours of Greece

Due date for project___________________________

Part A.

Complete the following brief fact sheet of Greece today to give to your tourists. You will find the information you need in a world almanac.

Government

Population

Area

Religion

Economy

Major cities

Climate

Literacy rate

Men

Women

Name________________
Date________________

Part B.

On your own paper, draw or trace a map of the country. On it, locate the following places.

Pindus Mountains

Mt. Olympus

Island of Crete and Rhodes

Peloponnesus (southern peninsula)

Mediterranean, Ionian, and Aegean seas

Athens and Thessoloniki

Other points of interest featured on your tour

Part C.

Prepare your itinerary for an eight-day tour of Greece and one or two of its islands. The first and last days have been done for you to help you get started. On your own paper, complete the itinerary for days 2–7.

Day 1 On our first full day in Athens, we take a full-day bus tour of the city to get a sense of our new surroundings. We hope for a beautiful, bright sunny day with the temperature in the 50s. Our tour bus, after driving us around to view the sights of the modern capital and Greece's largest city, suddenly makes a U-turn and you see down the side street the tall columns of the Parthenon built in the fifth century B.C., towering over the traffic, the modern shops, and high-rise office buildings. When the bus stops we climb to the top of the Acropolis where we finally get a closer look at one of the world's most renowned monuments of classical Greek architecture. When we come down, we have lunch at a small cafe and visit the National Museum where statues and parts of other antiquities are preserved. Back at our hotel, we take a quick nap before dinner at 8 P.M. We go to a local taverna where we enjoy a couple of hours of traditional and modern Greek singing, dancing, food, and fun before returning to our hotel and heading straight to bed.

Day 2

Day 3

Day 4

Day 5

Day 6

Day 7

Day 8 Our last drive from Athens takes us to its modern airport where we board Olympic Airlines for our long flight home.

Part D.

On your own paper, design a travel brochure to introduce travelers to the delights awaiting those who sign up for your tour. You may wish to include a map and picture of representative places featured on your tour.

Lesson 21
U.S.S.R. Dissolution—The Death of the Great Russian Bear

Objectives

- To examine the change from the Union of Soviet Socialist Republics (U.S.S.R.) to the Republic of Russia
- To identify geographical features of Russia and their effects on human-environment interaction
- To investigate key issues relevant to the "new" Russia

Notes to the Teacher

This lesson helps students bridge the transition from the "old" Russia (the U.S.S.R.) to the "new" Russia (the Republic of Russia). The history of Russia under the Tsar and of the Soviet Union under the Communists is presented in a brief reading. The economic and geographic features of Russia and its neighbors in the Commonwealth of Independent States are also briefly outlined in the reading selection. With this thumbnail sketch of Russia's evolution to its current status, students should have some understanding of Russia's background. Supplement the background material as necessary.

Students read and discuss the background material contained in the reading and examine a wall map that shows the new independent states. Climate and physical features of the area are located and discussed. Working in small groups, students deal with issues important in Russia today. They use their creative ideas in problem-solving situations.

Procedure

1. Distribute and have students read **Handout 54**. Locate pictures of Tsar Nicholas and his family, Lenin, Stalin, Gorbachev, Yeltsin, Putin, pictures of the Kremlin, Moscow, Leningrad, Siberia, etc.
2. Show the pictures of famous Russian people and places. Discuss the climate of Russia and how it affects people and their lifestyles.

   Some possible discussion questions:

   - Why do you think a bear was used to symbolize the Soviet Union? *(Possibly because of a bear's great strength—like a bear, the Soviets took over many of the smaller, surrounding countries and forced them to become part of the U.S.S.R.)*
   - If you were living in Russia today, would you be able to find a job easily? *(No, because today not everyone is guaranteed a job as they were under communism. Also, Russians have changed the way they conduct business, and, therefore, there are fewer available jobs.)*
   - When we talk about human rights, what do we mean and how has that changed Russia? *(Human rights are the freedoms we deserve as a people—the freedoms of speech, religion, press, the right to assemble, etc. Russians now can have those freedoms, and many people are pleased to be able to attend church openly and without fear of interference and to read about things that have happened without government censorship, etc.)*
3. Divide the class into small groups of four or five students. Distribute **Handout 55** to them and assign one issue to each group. Make sure students have large poster board, markers, an atlas, and an outline map of the Commonwealth of Independent States.
4. Allow students ample time to complete the activity. Have them present their posters and speeches to the class. Help students evaluate themselves and each other by providing a rubric which can be a grading tool.

Sample Rubric

Presenter's Name ____________________

Did the presenter....?

	Yes/No
Have good posture	______
Look at the audience	______
Speak clearly	______
Keep to the time limit	______
Follow the correct report format	______
Give correct information	______
Cover the subject thoroughly	______
Use visuals	______

For each yes response, give the presenter ___ points. For each no response, give a zero. Total the points and place here: ______

5. Use the students' completed work as a classroom (or larger audience area) display.

Enrichment/Extension

1. Research some of the early tsars. Prepare a two to three page biography of the tsar of your choice.
2. Research the cities of Moscow and Leningrad. Compare them to cities in the United States like New York City or Los Angeles. Explain how geography affects each of these cities.
3. Find out about cultural influences Russia has had upon the United States, *e.g.*, ballet, literature, etc., and write a paper on the topic to share with classmates.

Name________________
Date________________

The Death of the Great Russian Bear

Read the following selection carefully. Think about the many changes Russia has experienced. Be prepared to discuss the reading in class.

What happened to the Great Bear that once was the Soviet Union? The old tsarist Russia was destroyed by the October Revolution of 1917. Following a civil war, a communist government was established in 1922. Many of the poor people sided with Lenin and his Communist revolutionaries, and Russia fell. Under the new Communist leadership many small neighboring nations (Estonia, Latvia, Belorussia, Lithuania, etc.) were "eaten up" by the Great Russian Bear, and they became a part of the Union of Soviet Socialist Republics—the U.S.S.R.

With Communism, a better life was promised to those who lived in the Soviet Union. Everyone had a job and everyone had a vote. Of course, there was only one person to vote for and the people who got the really good jobs were those who belonged to the Communist party. People lived in crowded conditions, money was in short supply, and there were long lines for food and other goods that one wished to purchase. Most of the nation's resources went into the production of defense items with little effort made toward producing washing machines, automobiles, and other consumer goods. Everything was censored by the government before one could read it or view it.

People in the Soviet Union began to grow restless and eventually their leaders began to listen to their complaints. They knew that times were changing and they had to change as well. Censorship began to slow down and complete government control over their economic lives stopped. People could buy and sell freely now. The government recognized their human rights, and people were now allowed to practice their religion openly for the first time in many years.

In July, 1990, the Republic of Russia declared itself to be independent of the Soviet Union. Boris Yeltsin was elected its first president on June 12, 1991. A new era had begun. Twelve of the original fifteen republics in the old Soviet Union joined together and became the Commonwealth of Independent States (C.I.S.). These independent states had killed the Great Russian Bear! Yeltsin was succeeded by Vladimir Putin on December 31, 1999.

The Republic of Russia covers more than 6.5 million square miles extending from the Arctic Ocean in the north to the Black Sea in the south and east from the Baltic Sea to the Pacific Ocean. The Republic of Russia stretches across two continents—Europe and Asia. There are more than ninety different ethnic groups and nationalities living there and eighty different languages are spoken.

The Ural Mountains separate the Russian republic with Europe to the east and Asia to the west. A large plain lies on either side of the Urals, and a plateau area occupies the middle of the Siberian area. There are several large rivers that cross Russia that serve to drain the area. Many of them flow northward into seas that are frozen most of the year.

Most of Russia's climate is a continental climate which means that it has very cold winters and very hot summers. There are no large oceans or other bodies of water to help moderate the temperature. The northernmost part of Russia has a polar climate. Some of the land extends beyond the Arctic Circle. This tundra contains a permafrost layer (permanently frozen ground) except right along the coast where the water moderates the temperature enough to cause it to be a thinner layer of permafrost layer than is found in the interior.

Name________________
Date________________

Russia is rich in natural resources. There is plenty of oil, gas, lumber, gold, and fish. Before its independence, Russia supplied more than three-fourths of the Soviet Union's natural gas and 90 percent of its oil. The other states in the C.I.S. have to buy their oil and gas from Russia today. It used to be fairly cheap in the old days of the Soviet Union. Even though the price is greater now, the Russians have tried to keep the prices as low as possible so that the other states will remain in the C.I.S. In fact, Russia could be totally independent if it chose since 75 percent of the Old Soviet Union's industry is located within its borders. Russia would like all the states in C.I.S. to trade with one another so it buys grain from the states located on the Baltic. Russia supplies most of the manufactured goods needed by the other states in the C.I.S.

Most of Russia's people live in the western portion of the country and mainly around Moscow, the capital. Siberia has very few people and in some of the eastern areas of Siberia there are none. Most people live near and travel by train in this Siberian region. The Trans-Siberian Railroad connects western Russia with eastern Russia. A new railroad built to the north of the Trans-Siberian Railroad, the Baikal-Amur-Mainline (B.A.M.), and running parallel with it was placed there in hope that people would come to the empty Siberian countryside to find and to develop resources such as iron ore, gold, uranium, diamonds, coal, and lumber that are believed to be there.

All of Russia's large cities and most of its farmland and industry are found in the west. All of the government offices are located here, so all decisions made about the country are made in its western half. Russia is a country of two faces—the western one is filled with people and dwindling resources, and the eastern one has few people and many resources. Their challenge is to link the two regions to make one productive country—Russia.

Name________________
Date________________

The Death of the Great Russian Bear

Listed below are four scenarios. Read the one assigned to your group and follow all the directions. Be sure to add color to your poster creations and make them as neat and attractive as you can. Each group must do a map.

1. **Stop Polluting Lake Baikal!!!**

   The "blue eye of Siberia," as Lake Baikal is known, is the biggest lake in Russia. It is also the oldest and deepest lake in the world, and it holds 20 percent of the world's fresh water. It was known for its purity because of a little crustacean that eats the bacteria that would have caused impurities. Sadly, however, a plant built on the southern end of the lake in 1961 has been polluting it. The pollution area covers a thirty-mile range. Since Russia's independence, people have heard the news of this pollution. They are angry! Factories along the Selenge River, which feeds Lake Baikal, are polluting the river. A rare Siberian salmon breeds in this river. In the old Soviet Union, there were no laws against pollution. Your task is as follows:

   - On an outline map of Russia, locate and label Lake Baikal and its river tributaries.
   - Make posters to support the issue of a clean Lake Baikal.
   - Write a law for the Republic of Russia concerning pollution. Present it to the class.

2. **Come Work in Siberia!!**

   There are many resources in Siberia. Workers are needed to search for and to mine these minerals. Russia is rich in resources, but most people do not want to live in these cold regions. Your task is as follows:

   - On an outline map of Russia, locate and label Russia's industries and mineral resources.
   - Create posters to recruit people to come to Siberia to work. Be positive. Make the job inviting.
   - Prepare a speech to persuade people to come to Siberia for jobs. Deliver it to the class.

3. **Economic Interdependence among Russia and the C.I.S.**

   Even though Russia is rich in natural resources and most of the old Soviet Union's industries were located within its borders, Russia would like to be trading partners with the other states in the C.I.S. Russia buys things from the other states that they could make themselves, but they choose to buy from the C.I.S. They export goods to the C.I.S., especially manufactured goods. To keep these other states trading with Russia, your group has been put in charge of an advertising campaign to promote Russian goods. Your task is as follows:

   - On an outline map of Russia, locate and label Russia and the fifteen independent countries that were formerly the Soviet Union. Color (a single color) the twelve countries that formed the C.I.S.
   - Create posters with advertising information that would persuade the C.I.S. countries to trade with Russia.
   - Prepare a speech to be given to C.I.S. economic advisers to persuade them to be trading partners with Russia. Deliver your speech to the class.

Name__________________
Date__________________

4. **Be an Entrepreneur in Russia**

Now that Russia is engaging in the free enterprise system people can start their own businesses and become entrepreneurs. Entrepreneurs are people who have an idea for a new business, get financial backing, and take the risk involved in starting a new business. Russia is a country that does not have many of the consumer goods and services that we have in the United States. They do not have a lot of money to spend these days. Your task is as follows:

- On an outline map of Russia, locate and label Russia's largest cities, rivers, and railroads.
- Decide upon a business you would like to operate in Russia. Write a business plan for your business. How much money might you need to get started? Where would you get this money? What kind of equipment and supplies would be needed? What price would you charge for your goods or services?
- Create posters to advertise for Russian workers to come work for you. Make the ads inviting so that you can hire good workers. Present your business plan to the class.

Lesson 22
Asia's Industrial Giant—Made in Japan

Objectives

- To examine the relationship of Japan's geography to its demographics and economics
- To identify Japan's economic strengths and weaknesses
- To compare Japan's industrial system to that of the United States

Notes to the Teacher

This lesson examines Japan, a non-western country, to explore its physical and cultural geography and determine how Japan's geography is linked to its economic growth. Japan has moved up in its ranking as an industrial nation since 1950 when it held no ranking because it did not export anything. Today it is highly competitive against superpowers like the United States and European countries like Germany.

It is important that students have access to atlases and other information about Japan and the United States in order to complete their assignments. Research time will be needed.

Students complete research and compare Japan with the United States in the areas like geography, demographics, and economics. Students complete a personal survey of products and their origin. Students then examine the role of the Japanese worker in a company and compare it to that of a U.S. worker.

Procedure

1. Ask the class, "What do you know about what you own or use at home?" (Brand names may be given.) Tell students that they are being sent on a mission. Give them **Handout 56** and instruct them to take it home and complete it by finding where the item was made and why that particular brand was purchased.

2. Tally the student survey results and discuss the conclusions that can be drawn.

   Suggested Responses:

   1. *Most of the items will come from Asia, with Japan being a primary country.*

   2. *Single out Japan by talking about the events following World War II when Japan was under U.S. control and how Japan modernized and westernized its industry and, to some degree, its culture.*

3. Distribute **Handout 57** and have students complete it using an atlas and other research material. Consider having students work in cooperative learning groups.

4. Debrief the completed charts. Ask students why people in the United States do not live as closely together. How has the lack of flat land affected the lifestyle of the Japanese? With no natural resources, how have the Japanese become so successful as a leading industrial nation?

Suggested Responses:

Japan

Population	**Land area per square mile**	**Industries**	**Agricultural products**	**Minerals**
125,107,000 **Population density** 857 per square mile	145,883	Electrical Electrical equipment Autos Machinery Chemicals Fishing Shipbuilding	Rice Grains Vegetables Fruits	none

United States				
Population	**Land area per square mile**	**Industries**	**Agricultural products**	**Minerals**
260,714,000	3,787,318	Manufacturing: lumbering and products	Dairy	Aluminum (Bauxite)
Population Density		Furniture and fixtures	Cattle	Copper
70.3 per square mile		Clay, glass, and stone products	Poultry	Iron
		Iron and steel	Hogs	Gold
		Machinery	Corn	Lead
		Computer equipment	Soybeans	Nickel
		Electrical machinery	Wheat	Tin
		Motor vehicles and parts	Vegetables	Silver
		Textiles	Fruit	Uranium
		Tobacco products	Nuts	Zinc
		Clothing	Cotton	Oil
		Paper and paper products	Tobacco	Natural Gas
		Printing and publishing	Mixed farming	Coal
		Chemicals		
		Petroleum products		
		Rubber and plastic products		

California				
Population 31,210,750 **Population density** 190.8 per square mile	**Land area per square mile** 163,707	**Industries** Agriculture Manufacturing: Foods, printed materials Primary and fabricated metals Machinery Electric and electronic equipment Transportation equipment Instruments Commercial fishing	**Agricultural products** Grapes Cotton Lettuce Flowers Oranges Nursery products Walnuts Hay Strawberries Almonds Broccoli Sugar Beets Peaches Potatoes Livestock	**Minerals** Sand Gravel Boron Cement Diatomite Asbestos Calcined-Gypsum

Indiana				
Population 5,712,779 **Population density** 154.6 per square mile	**Land area per square mile** 36,420	**Industries** Manufacturing: Primary metals, transportation equipment Industrial machinery and equipment Electronic and electric equipment Services Agriculture Government Wholesale and retail trade Transportation and public utilities	**Agricultural products** Corn Wheat Soybeans Nursery and greenhouse products Vegetables Sweet corn Melons Hay Livestock	**Minerals** Crushed stone Cement Sand Gravel

5. Distribute **Handout 58** and allow students to read it. Take students back to the questions they discussed when using the chart and help them see if their answers match or if the new information in the handout helped them formulate new answers.

6. Use **Handout 59** to determine the work ethic/cultural traits of Japanese workers and to compare it with what students know about U.S. factory workers. After reading the scenario, and using information from the previous handout, complete the handout in groups. Have students discuss whether or not they feel that the Japanese factory system would work in this country and what changes do they feel might be necessary in the United States to make it work? Allow students plenty of time and opportunity to express themselves on these issues. If hostility towards Japanese products and processes exist, try to get students to look at their feelings objectively—use facts not feelings. (Some students have parents who work in businesses in competition with Japanese-made products. Their emotions may run strong.)

Suggested Responses:

Cultural Traits—Japanese workers	How is it evident/or not?	Cultural traits—United States workers
Sense of duty to company and/or society	Loyal to company, often for life; family feeling	Moves to a new job if working conditions and/or pay is better elsewhere
Individuality defined in a group context	Workers see themselves as part of a group and work together	Workers see themselves as individuals even when working on assembly lines
Unified culture	Few foreigners in Japan	Many cultures represented in United States
Thrift and hard work	Workers work hard for corporations in return for small wages and benefits	Workers work hard; many save money
Respect for education	Most place a high value on education	Some workers place high value on education
Adaptability	Ready to retrain for new job when old job becomes obsolete	Some workers ready to retrain for a new job
Imitators, then innovators	Create new high-tech products from those already on market—electronic, computers, robots, etc.	United States has always been known for its inventiveness

Enrichment/Extension

1. Research Japanese businesses in Japan. Find out how they operate with regard to their employees. What benefits do they receive? When these companies (*e.g.*, Toyota, Honda) open factories in the United States, what changes must be made in their approach to the employees? Write an essay on your findings and share it with the class.

2. Investigate the relationship between Japan and the United States with regard to trade. What problems have they had? What are the biggest areas of controversy? What steps have been taken to solve their problems? What predictions can be made about a positive trade relationship between the two countries in the future? Write an essay on your findings and share it with the class.

Name__________________
Date__________________

A Survey—Where Was It Made?

Record the name of the country where each of the following items that you own were made. Explain briefly why a particular brand of an item was chosen—for example, was it quality, price, availability, need, etc.? After completing the chart, answer the questions.

Item	**Country of origin**	**Why this item (brand) was purchased**
Television		
Walkman radio		
VCR		
Stereo equipment		
Computer		
Camera		
Video game machine		
Video game tapes		
Camcorder		
Film (snapshots, video, etc.)		
CD player		

1. Which country dominates the others in the production of these items?

2. What conclusions can you draw from your list about certain countries and their industries?

Name________________
Date________________

The United States and Japan—A Comparison

Use an atlas, almanac, or encyclopedia to complete the chart.

	Population/ population density	**Land area per square mile**	**Industries**	**Agricultural products**	**Minerals**
Japan					
United States					
California					
Indiana					

Name________________
Date________________

Made in Japan

Read the following information about Japan. Then revisit the questions you answered about the chart on the previous handout: Why don't people in the U.S. live closer together? How did the lack of flat land affect the lifestyle of the Japanese? Without natural resources, how have the Japanese become successful as a leading industrial nation? Have your views changed after reading this information?

Japan is an archipelago—that means that it is a group of islands. It has four main islands and about 3000 smaller ones with a total area of 145,850 square miles. About three-fourths of Japan is covered with mountains. Many of these mountains are volcanic—67 of them are still active. The most famous mountain is Mount Fuji, Japan's highest mountain. Japan has many earthquakes—sometimes as many as four a day. Some are so slight that one would hardly notice; others are more severe. Very little of Japan is level—only about 15 percent of its land is fit for farming. Some plains can be found along the coast. Tokyo, Japan's capital, is located on one of these coastal plains.

Japan has a wide range of climates because it is located between 24° N latitude and 45° N latitude. In the north summers are short, cool, and damp, and winters are long and cold. In the south the temperatures range from 50°F to 90°F. Monsoon winds cause great differences in the temperature and precipitation. Ocean currents also affect Japan's climate—tropical currents warm the land and cold, Arctic currents cause a cooling effect. Dangerous fogs occur when the two currents meet.

Japan's population is 125,107,000 with a population density of 857 people per square mile. Most people live in the small plains area because Japan is so mountainous; therefore, people must live very close together in small villages or towns providing for their needs by farming or fishing. Tokyo is considered to be one of the largest cities in the world with a population of about 30 million people. Most of the people who live here are Japanese (99.4 percent) with a small (0.6 percent) Korean population. Japanese is the principal language. The majority of people belong to the religions of Buddhism or Shintoism. Japan has a parliamentary democracy with an emperor as head of state and a prime minister as head of the government. The unit of money is the yen.

Japan is one of the world's most important industrial nations, but it produces its goods with imported resources because it has none of its own. Japan must import oil and coal to make the energy needed to run its industries as well.

Until 1950, the Japanese produced most of their own food, textiles, machinery, steel, and vehicles, but as they began to Westernize and modernize their industries, they were able to export many of these goods. In fact they are now the major competitor for United States and European companies. The first industry affected by Japan's competition was shipbuilding with motorcycles, cars, vehicles of various types, cameras, electrical goods, electronic equipment, and musical instruments to follow. Japan is now involved with the production of all kinds of industrial goods. The Japanese invest large amounts of money in their businesses to keep them modern, and their management system has been successful. With these and their loyal employees, they are able to produce high quality goods to sell at home and abroad.

Name________________
Date________________

Working in a Japanese company is different than working for a U.S. company. Most Japanese work for one of just a few large companies that are run by families—like Toyota or Mitsubishi. When one works for these companies, one is treated as a family member. The company takes care of its workers for the rest of their lives. Sometimes workers even live in a small city set up for employees where the cost of housing is less than it would be elsewhere. Some companies make sure that scholarships are given to the children of their employees. Health care is provided for employees and their families. Vacations and big parties are usually supplied each year. In the factories, there is a team spirit, and the managers work alongside the laborers with little difference being made for their higher position. People like to work under these conditions and give the company their complete loyalty. There are no strikes.

Name________________
Date________________

Made in Japan—Portrait of Japanese Workers

The column on the left lists cultural traits of the Japanese people. In the second column, tell how you can identify that trait from this reading or information found in the last handout. In the last column, compare the traits with U.S. workers.

The Japanese economy is dominated by large corporations. These corporations are often run by *Zaibatsu*, financial groups who are the leaders of modern Japanese industry. The *Zaibatsu* import raw materials, assemble or build products in Japan, and export them throughout the world. Lately, they have shifted their emphasis from heavy industry, like steel production, to light industry, such as computers and electronics, or they have changed a labor intensive heavy industry, like automobile production, to high-tech production using robot assembly.

The people who work for the large corporations see them as paternalistic and fatherly. Workers are very loyal to the company often working for the same corporation for their entire lives. Employees often have lifetime security in their jobs, and if a job becomes obsolete, a new job is provided along with any retraining that is needed. Their salaries are modest, but the fringe benefits make up for it. Sometimes housing is provided or at reduced cost. Cafeterias, daycare facilities, free travel vacations, etc., are given to the workers.

The attitude of the Japanese worker in the corporation is the same as being part of a family or a club. Japanese workers believe that when everyone does their best, everyone benefits from a job well done.

Cultural traits—Japanese workers	**How is it evident/or not?**	**Cultural traits—United States workers**
Sense of duty to company and/ or society		
Individuality defined in a group context		
Unified culture		
Thrift and hard work		
Respect for education		
Adaptability		
Imitators, then innovators		

Lesson 23
China—An Ethnic Quilt

Objectives

- To locate some of China's ethnic minorities on a map of the country
- To recognize the cultural individuality of China's ethnic minorities

Notes to the Teacher

Western society often regards China stereotypically as a nation of mass conformity and cultural sameness. This model is based upon the majority population, the Han, who comprise 92 percent of China's people. The ninety-five million people who make up the national minorities inhabit more than half of China's territory. These minority groups differ from the dominant Han Chinese in geography, language, customs, religion, race, and historical development. Communication between the Han majority and the national minorities works through one standard language, Mandarin (which 68 percent of Chinese understand), despite the country's seven major dialects, and through written picture words. However, picture words are difficult to learn; a person with knowledge of 3,000 to 5,000 picture words out of 40,000 is considered literate.

Chinese culture is 7,000 years old, making it perhaps the world's most ancient continuous culture. Twenty-eight percent of its 1.2 billion people live in urban areas; most of the rest of the people are engaged in agricultural work in rural areas, although some of its people are still nomadic herdsmen. Population density varies from 1,000 + per square mile (Kiangsu) to 3 per square mile (areas in Tibet). Religious beliefs are varied, from indigenous animism, Confucianism and Taoism, to Buddhism, Islam, and Tibetan Lamaism (Buddhist).

Traditionally dynastic, China founded a Republic in 1912 which was overthrown in 1949 when Communism became the Chinese form of government. The Communists promised to protect the national minorities in China by granting autonomy in the ethnic regions and providing assistance in maintaining languages and cultural identities. Over 400 ethnic groups registered, and the government currently recognizes fifty-five national minorities, with many others under consideration. Minorities are exempted from the birth control programs required of the Han majority.

Students locate ten of China's ethnic groups on a map. They then interpret and compare two Chinese legends to learn more about the country's cultural diversity.

This lesson may take two days to complete. The lesson serves to illustrate that other nations share an ethnic diversity much like our own with all the richness—and problems—that such variety brings.

Procedure

1. Begin by asking students to share what they know about modern China. (In most cases, students present a stereotypical preconception of China and its people: overcrowding, mass conformity, sameness of culture.) After students have shared this stereotype, inform them that they have given a stereotype of eastern China, and of the dominant Han culture. It may help to review stereotypes from Lesson 6, on ethnocentricity. Move then to the focus of the lesson, the 8 percent ethnic minority that populates the western half of China. Share information with the class from the

	China	**United States**
Area	3,696,100 square miles	3,787,318 square miles
Population	1,190,430,000	260,714,000
Cultural Age	7,000 years	400 years
Location	21° N–53° N latitude 73° E–135° E longitude	24° N–49° N latitude 66° W–124° W longitude

Notes to the Teacher to supplement material on modern China from the text. It may be useful to make a transparency or put the comparison chart at the bottom of page 143 on the chalkboard to illustrate similarities as well as the vast differences between China and the United States.

2. Distribute **Handout 60**. For part A, either allow students time to read the descriptions of the ten Chinese ethnic groups silently, or have students volunteer to read different ones aloud to the class until all ten are introduced. Part B can be completed either individually or as a paired activity.

3. **Handout 60**, part B can be enriched by having students assign colors to each of the ethnic groups and then coloring in the map, creating a color quilt of ethnic groups that both visually reinforces the point of the handout, and provides a physical activity for a "hands-on" student. Additionally, part A can be cut up into paper rectangles showing each ethnic group, connected to the map in part B by colored yarn of the same color as used on the colored map in an attractive bulletin board display.

Suggested Responses, Part B:

1. *Han*
2. *Zhuang*
3. *Miao*
4. *Yi*
5. *Tibetan*
6. *Di*
7. *Yugur*
8. *Tu*
9. *Mui*
10. *Mongol*

4. Distribute **Handouts 61** and **62**. Allow students time to read the selections and answer the questions which follow. Consider dividing the class and giving each half one of the two handouts. Have students share their responses to provide information feedback on location of information and speculative information.

Suggested Responses, Handout 61:

1. Water spiders

2. When they swallowed the water, they also swallowed the written characters

3. Perhaps a migration or a flood

Suggested Responses, Handout 62:

1. For defense

2. To keep away flies

3. The Di would sing to keep it awake.

4. Hygiene—combing hair once a day; healthy diet—eat three times a day; mental attitude—wing while working to make it pleasant

5. Perhaps learning to care for domestic animals, or from a time of food shortage

5. Conclude by comparing the two legends. Ask what the two legends seem to have in common. (*They may both have derived from actual historical origins; both show the relationship of humans and other living creatures; both help to establish individual cultural identities of the ethnic groups.*) See if students can determine what the entire lesson has contributed to their perception of ethnic minorities in China. (*Answers may include the idea that China has more diversity than it is stereotypically known for, and that each minority has its own geographic and cultural individuality.*) The following might be a good culminating question: Do you think that what you found true in China relative to its minorities is true in other countries also? Why or why not? (*Most will respond yes. Many cultures have subcultures and few countries have only one dominant culture and no others.*)

Enrichment/Extension

1. Make a graph or chart that shows a comparison of the ten ethnic groups introduced in this lesson. Due to the large number of Han—92 percent—you may wish to chart just the nine minority groups.

2. Trace the outline of both China and the United States on the same map scale—showing by overlay or superimposing how China and the United States are similar in both size and latitudinal and longitudinal locations.

Name________________
Date________________

Meet China's Ethnic Groups

Part A.

Read the descriptions below which identify ten of China's ethnic groups.

1. **Han** The Han are the majority group in China, making up 92 percent of the population. The name comes from the dominant dynasty, or ruling family, in the third century B.C. They live in the east and the northeast.

2. **Zhuang or Chuang** The Zhuang are the largest minority in China. Their population lives primarily in south central China. They share language similarities as well as festival singing and dancing customs with the people of Thailand to the south.

3. **Miao** The Miao have no written language. They are known for their craftsmanship in embroidery. These superstitious mountain people are related to the Hmong tribes found in neighboring Vietnam, Laos, and Kampuchea to the south.

4. **Yi** The language of the Yi is a synthesis of Tibetan and Burmese. Traditionally a society of fierce warriors, they maintained until the 1950s a slave society in which even slaves could own slaves. They have a religion based upon studies of sacred Buddhist writings.

5. **Tibetan** The Tibetans inhabit the mountains around the Himalaya Mountains which separate China from India. Mostly farmers and herdsmen, they practice Lamaistic Buddhism. Tibet did not come under full Chinese control until 1959.

6. **Di** The Di are a small subgroup of Tibetans known as the White Horse Tibetans. They appeared in ancient Chinese histories before 420 A.D. Although they have no written language, they do have a colorful oral history.

7. **Yugur or Uigher** The Yugurs are basically a farming culture. They speak a language that is related to Turkish. Religiously, they practice the Islamic faith.

8. **Tu** The Tu are former nomads related to the Mongols. They are noted for traditional ceremonies and sense of hospitality. Historically, they serve as frontier guards for Imperial China.

9. **Hui** The Hui who live in north central China are distinguished from the Han majority only in that they practice the Islamic religion.

10. **Mongols** The Mongols who live on the north central frontier are essentially farmers. Semi-nomadic live-stockmen among them still live in felt tents called *yurts*. Many practice the Lamaist religion of Tibet.

Name____________________
Date____________________

China's Ethnic Quilt

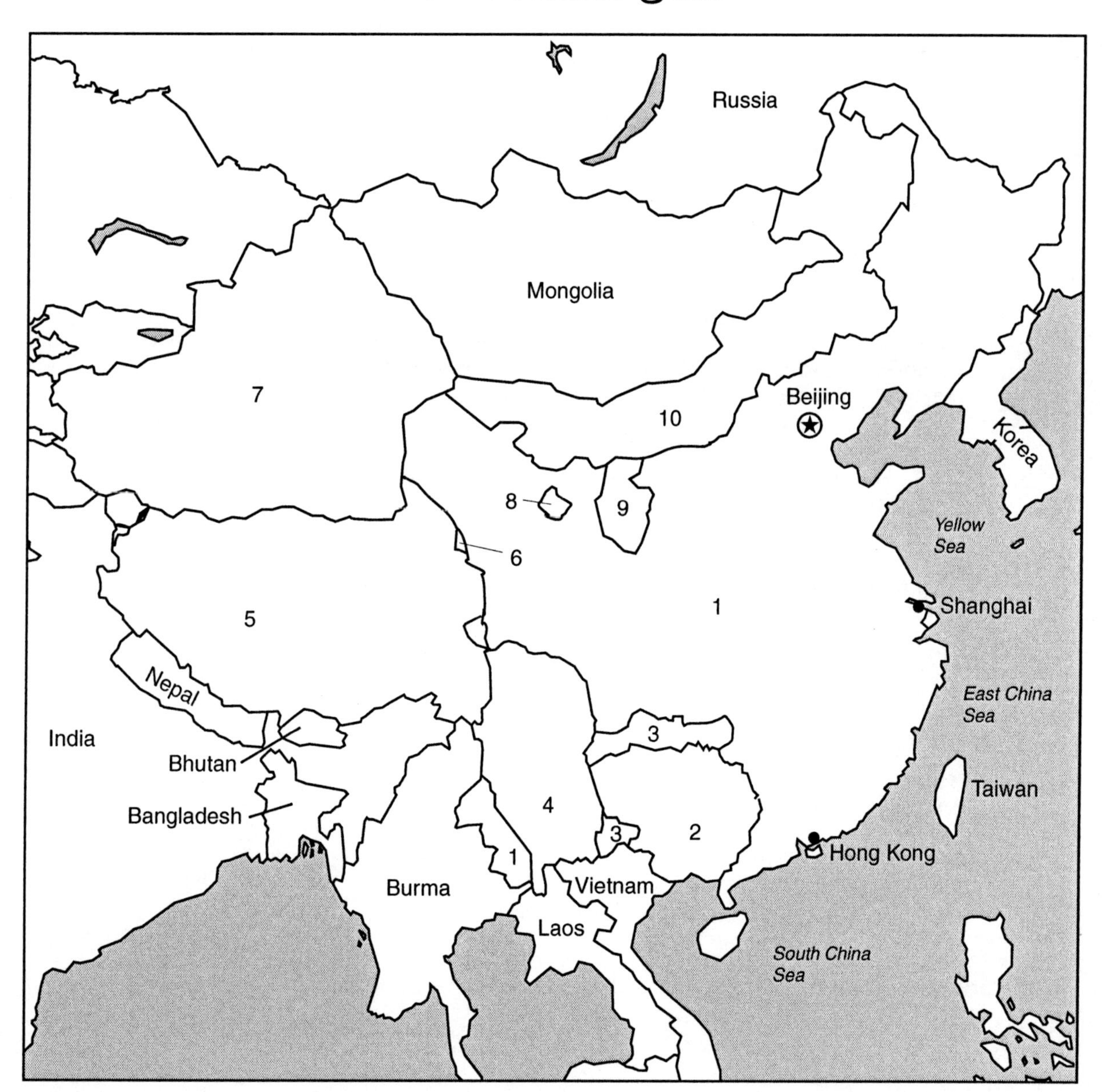

Part B.

Using the identifications in part A, locate the corresponding number on the above map which shows the geographic location of each ethnic group. As you locate them, write the name of the ethnic group after the correct number below.

1.
2.
3.
4.
5.
6.
7.
8.
9.
10.

Name________________
Date________________

Why the Miao Have No Written Language

Read the legend which follows, then answer the questions. Be prepared for class discussion.

> The central and southern parts of Guizhou are inhabited by the Hei Miao, who are by far the largest group of Miao in the province. According to legend, the first Miao came from Jiangxi Province to the east. Another legend explains why the Miao do not have a written language.
>
> Long, long ago, runs the legend, the Miao lived in the same area as the Han Chinese. But the Han were too cunning for the Miao, so the latter decided to move away. They traveled a great distance and finally came to a broad stretch of water that, as they had no boats, they were unable to cross.
>
> At that time, according to the legend, the Miao knew at least some written characters. As they stood beside the water pondering what to do, they noticed the water spiders walking about on the surface. They said to themselves, "If these little things can walk on water, why can't we?"
>
> So the Miao tried, with the result that they nearly drowned. In struggling to regain the bank they swallowed a lot of water, and along with the water they swallowed all the written characters they knew. Since that day the Miao have been without any written language.[1]

1. How did the Miao get the idea to try to walk on the water?

2. How did the Miao lose the written language that they once had?

3. What actual or historical cause may have been the origin for this legend?

[1]Wong How-Man, "Peoples of China's Far Provinces," *National Geographic*, (March 1984): 305.

Name________________
Date________________

Why the Di Are Always Singing

Read the legend which follows, then answer the questions.

> In ancient times, runs the legend, heaven bestowed on humans an abundance of rice covering the entire earth like snow. But a woman accidentally stepped on some grains of rice, thereby offending God. God sent the ox to earth to announce His punishment for mankind: Each person was to comb his or her hair three times a day and eat but one meal a day.
>
> By mistake the ox ordered combing of the hair once a day and eating of three meals a day. God was much angered and banished the ox to earth to toil and repent. The ox begged for mercy. First, he claimed he would be ill-treated on earth. God therefore gave the ox horns to defend himself. Second, the ox worried about insects bites. God gave him a tail to drive the insects away. Third, the ox was afraid of being punished if he was to oversleep. So God asked the people to sing to the ox to keep him awake. To this day, the Di always sing whenever they plow the fields with their oxen.[2]

1. Why was the ox given horns?

2. Why was the ox given a tail?

3. How would the ox avoid oversleeping?

4. What essentials of good health are taught in this legend?

5. What actual or historical cause may have been the origin of this legend?

[2]Ibid, 320–321.

Lesson 24
Singapore—A City of Many Faces

Objectives

- To examine the geographic, economic, and environmental conditions of the island of Singapore
- To practice map skills by enlarging a map of Singapore using a grid pattern

Notes to the Teacher

This lesson examines some of the main features of the little island country of Singapore. Its geography, economics, and environment are some of the features that have allowed it to become prosperous. Singapore's history as a part of the British Empire and its move to independence is at the heart of a story that created an island economy that is interdependent in the world of international trade and has made it one of the few southeastern Asian countries that has become financially successful. Singapore's climate could have been a detriment because of the heat and high humidity, yet it has overcome this feature, and many people come here for employment.

Students read and discuss information about the island and do some research about the area. Students explore the relationship of the people with their environment. Finally students complete a map activity that asks them to enlarge a map of the island by using a grid system.

Procedure

1. Place an umbrella, a fan, and a toy lizard on a table. Ask students where in the world these three items would be something they needed almost daily. Allow them time for a discussion about where and why these things might be needed.
2. Use a world map to show the location of Singapore. Ask students to read **Handout 63**. Discuss what students have learned about the climate, geography, and environment of this island. Ask students to speculate about why the location of Singapore has helped to make it important in world trade. (It is located near Japan, Korea, and China, and some of the smaller southeastern Asian countries who produce trade goods for sale in the western hemisphere. Ships can stop here in its excellent port and unload goods heading in either direction.) Determine if students can now understand why the umbrella, lizard, and fan might be necessary items in Singapore?
3. Use atlases or books about Singapore to examine closely the island of Singapore. Talk about its absolute direction in terms of longitude and latitude (1° N, 104° E) and its relative location to the rest of the world. Make the connection between where it is located and how that benefits it economically in a trading situation.
4. Have students study **Handout 64**, part A, and note the grid system identified with letters and numbers and the shape of Singapore and the smaller islands that make up the country of Singapore. Tell students that on the larger grid paper, **Handout 64**, page 2, they are to redraw the map of Singapore so that it is larger. Help students understand that they must copy what is in the grid section B1 on the original handout so that on their blank grid section B1 looks the same—only larger.
5. After students have completed the new, larger maps of Singapore, have them work in groups with their research materials (atlases, encyclopedias, books, etc.) to label other features on the map. (These may include other cities, locations of forests, railroads, habitats of indigenous animals, etc.) Encourage students to make their maps colorful with colored pencils or markers. Make sure a legend is included that explains the added features that students chose to place on their maps.
6. Have students share their work. Use completed maps for a classroom display.

Enrichment/Extension

1. Conduct further research about Singapore's people. Make graphs showing the various percentages of ethnic groups that live and work in Singapore. Make other graphs and charts to illustrate the agricultural and manufactured goods produced on the island. Do additional graphics, *e.g.*, compare the amount of trade done in ten year segments between 1950 and 1990 to see the growth rate; track the growth of trade between Singapore and the United States from 1950 to 1995 in five year segments, etc.

2. Research the animals and/or plants that live on Singapore. Focus on any that are endangered and why that is the case. Write a report and use pictures of the animals and/or plants to tell about what is happening to them. Show the relationship between the human inhabitants and the other species on the island. Prepare a History-Day-type board display and use it to share your findings.

3. Write a report about Thomas Stamford Raffles, the British subject whose role in the East India Company caused him to single out Singapore as a place in which the company should invest its effort and money in 1819. Prepare a History-Day-type performance and present it to your classmates.

Name________________
Date________________

Singapore—A City, An Island, A Country

Read the following information very carefully.

When is a city also a country? Why does this place continually change its size and shape? The answer to both questions is: Singapore! It is a small island—only about one quarter the size of Long Island, New York. When it gets too crowded and more space is needed for building apartments and businesses, dirt is brought for landfill from nearby, tiny uninhabited islands. This landfill creates the necessary building space. The whole island of Singapore is only about twenty-six miles long and fourteen miles wide with an area of more than 238 square miles. Its capital city is Singapore, too, and it takes up most of the island. The name Singapore comes from the Malay word meaning Lion City.

Singapore is a small island country with an independent government. It is just off the tip of the Malay Peninsula and is close enough that it is linked to the peninsula by a causeway. The island has three distinct regions—in the middle are the highlands, in the west are the rolling lands, and in the east are the flat lands. The city of Singapore lies in the southern part of the island. The Singapore River is the most important river in the area. At one time everyone settled along the river and much traffic went up and down the river making it an important trading center. Today there is little traffic on the river except for small boats that might be found in the harbor at the river's mouth. The modern port is located between Singapore Island and Sentosa Island where the water is deeper and allows for larger vessels. This port is the second largest port in the world. Rotterdam in the Netherlands is the largest. The port of Singapore is growing at such a rapid rate that it may soon become number one!

Singapore is located just 87 miles north of the equator in the South China Sea. One has no need for seasonal clothing because there are no seasons! The temperature stays near 80°F all year. Since Singapore is an island it gets up to 100 inches of rain each year. The humidity remains so high that it is steamy. These are great conditions for tropical plants. Flowers and trees abound in and around the city. Some parts of the island are heavily forested. In order to make small farms in the countryside people must clear away this lush growth. Many species of plants and animals have become extinct on this island because of the high population growth and building programs that have taken away their habitat.

One reptile is a common house guest in a Singapore house. It is the house gecko. A gecko is a lizard who eats most of the insects in the home while the home gives it shelter from its enemies. It is an amiable relationship—one just has to be careful not to bother its eggs which may be found behind shelves or picture frames!

Singapore was basically a deserted island until the nineteenth century. Its only visitors were pirates who operated in the surrounding seas and the small settlement of about 150 people who lived by the river. It belonged to the Johor-Riau Sultanate. In 1819, Sir Stamford Raffles from Great Britain came to the island and helped the heir to the sultanate take control of this kingdom. In return Raffles was given the right to set up a trading post in Singapore.

A great trade developed, and the island prospered. Many people from China came to the island to work as well as people from the Malay peninsula. Indians and Thais came as well as Europeans. According to Raffles's plans for the building of the city of Singapore, there were areas of the city where each of these ethnic groups lived together. He believed that people who were of like cultures would be happier if they lived together. Even today, people

Name________________
Date________________

in Singapore are multicultural, and their government promotes the idea of getting along together. They do not appear to have many problems that concern their racial or ethnic identities. There are four official languages in Singapore—Malay, which is the national language; Tamil, which is the language spoken by most of the Indian population; English, which is the language in which business is conducted; and Mandarin (a dialect of Chinese), which is the language of the majority of the population. Most people speak many languages.

People in Singapore find work in the shipping and trade industries, manufacturing (machinery and appliances, electronics, clothes, paper goods, rubber and plastic items, chemicals, furniture, etc.) oil refining, land reclamation, construction, tourism, banking, and farming (poultry, pigs, tropical fruit, orchids, and alligators). Alligators are raised for their skins which are used in the manufacture of shoes and purses.

As you see, Singapore is a city with many faces. It is a city, an island, and a country.

Name________________
Date________________

Enlarging A Map

Part A.

Using part B, a 1¼-inch grid, copy this map. Draw each section of the map grid—one section at a time. Be careful to copy as nearly like the original as you can. For example, copy the part of the map that is in section B1 into the new grid section B1, then move into the next grid and copy what is in that grid. When you have finished you will have a new map that is larger than the first. Other features may then be added, and the map may be colored with colored pencils. *Be sure to make a legend if you add other features to your new map.*

Singapore

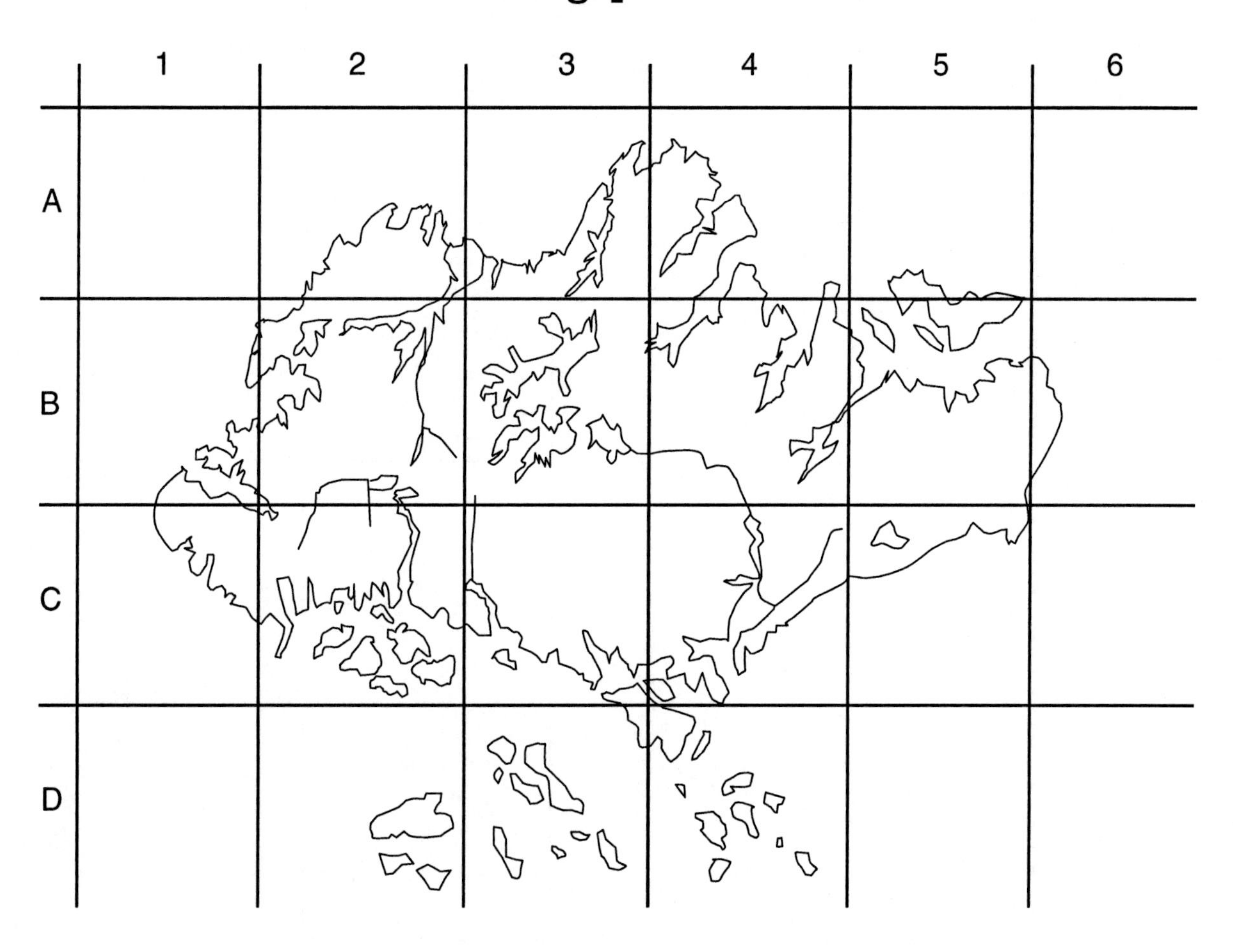

Name________________
Date________________

Part B.

	1	2	3	4	5	6
A						
B						
C						
D						

Lesson 25
India—Panorama of a Region

Objectives
- To examine India's geography as a region
- To identify human-environment interaction
- To research regional elements of India

Notes to the Teacher

India is a country is southern Asia and a member of the Commonwealth of Nations. India, once a member of the huge British Empire, is now a union of twenty-five states and seven territories. The Republic of India is governed under a constitution that incorporates various features of the constitutions found in the United States, Great Britain, and other Western democracies. It has a large population and many problems related to its population. Political unrest seems to be the normal pattern for the country with dissent based on religious and economic factors.

Students examine the definition of a region and the elements of the region to which people must adapt in order to survive. They use mapping, researching, and critical thinking skills to complete their assignment.

Students need access to atlases and research materials to complete a map of India showing important geographic and economic features. They discuss a reading about India and complete some related activities. Students research some features of India provided on a handout. The result is a large display, a panorama, of the region of India.

Procedure

1. On an overhead, display an outline of the country of India. See if students can identify it. Ask students when they identify India if they can name some of the neighboring countries or bodies of water. Then, use a globe and/or wall map and actually locate India and its neighboring countries or bodies of water.
2. Tell students that India is a region—a basic unit of geography. Regions are areas that have unifying characteristics. This is a way of organizing our knowledge of the world. Let them read about India on **Handout 65**.
3. Debrief the reading about India. Ask students some common characteristics of the region that could identify India as a geographic region (religion, physical features of the land, climate). Talk about ways people adapt to their environment (clothing, crops, housing, etc.). Discuss how Indians dress, eat, earn a living; where they build their homes and how they build them; how the irrigation system for dry areas works; and how they deal with the monsoons' heat and excessive rain—all examples of human-environment interaction.
4. Provide students with a map of India, **Handout 66**. Have students shade in the desert, plateau, and mountainous areas and label each one. Label major cities and rivers. Locate railroads and airports. Have students use an atlas to assist them in this. From the reading, label the areas where all the products mentioned in the reading are grown. Create a key for the symbols used. Consider putting students into groups to complete this assignment.
5. Divide students into groups of three to five. Distribute **Handout 67**. Allow students time to complete the assignment. Time is needed to research and to complete their illustrations. Be sure that students have poster board, markers, and tape.
6. Have each group present its completed panorama. Each groups' panorama should be different. Let students share the information they gathered. Display the completed panoramas.

Enrichment/Extension

1. Write to the Indian consulate in Washington, D.C. for information about India. Prepare and share a report based on the information they provide.
2. Invite a local person of Indian heritage to speak to the class and share knowledge about Indian culture. Ask the speaker to discuss language, religion, clothing, food, traditions, business, education, etc.

Name________________
Date________________

India—An Asian Region

Read the information in this article and locate some of the common characteristics that identify India as a region. Look for similarities in religion, physical features of the land, climate, etc. Notice how people adapt to their environment. Identify features of their culture, such as food, clothing, shelter, occupation, etc.

If you heard someone say that India is a subcontinent, what comes to mind? Is it under water? No! Is it a sandwich? No! Is it under a continent? No! No! No! Then, what is it? India is located on a subcontinent of Asia. (That means it is a large piece of land but smaller than a continent.) It sticks out into the Indian Ocean.

India's area is 1,222,559 square miles, which is about one-third the size of the United States. Its neighbors are Pakistan in the west; China, Nepal, and Bhutan in the north; and Myanmar (Burma) and Bangladesh in the east. The highest mountains in the world, the Himalayas, lie on India's northern boundary. Its climate varies. It is very hot, tropical, in the south but in the north it is extremely cold. The Assam Hills get about 400 inches of rain annually.

Europeans have been in contact with India for more than twenty centuries. During this period, India was not a united nation. It was, and still is, a region of much diversity because of its large number inhabitants.

Many languages are spoken and there are many geographical regions. There are also many religions, but the two largest religious groups are the Hindus and the Muslims.

Muslims believe that everyone should be given an equal opportunity to progress. Hindus use a caste system (classes) where people must remain in one level and cannot cross over into new areas. A person is born into a certain caste and must marry within that caste. The children must remain in the same caste as their parents. India has laws today against discrimination based upon caste, but they have not been very successful.

India is a crowded country. Its population is 1.03 billion, with the birthrate increasing each year. With better medical care, more infants survive and older people live longer. Many people have left their small villages to come to the cities to look for work. The cities are overcrowded and, because there are not enough jobs, many people go hungry and have to live in slum areas. The government has tried to encourage families to have only two children by giving those who limit their families special benefits.

India is a region of three seasons. The first season is cool and dry with clear skies, green landscape, and plenty of water. The temperature gradually rises as a hot dry season withers the vegetation and dries up the ponds. Finally, a shift in the winds causes monsoon rains which revive vegetation, and soon the landscape become green with summer crops. Sometimes the rains come too late for good crops. This results in famine and starvation for millions of India's people. The dangers of drought have lessened, however, as transportation has improved and irrigation has become widespread.

The highest mountain ranges in the world curve around India's northern boundary, but along its coast are river deltas only a few inches above sea level. In some parts of India there is too much water; in others, too little. The hills of the northeast have the heaviest rainfall in the world, averaging 500 inches a year, but in the northwest is one of the driest deserts in the world.

India is an agricultural land with the densest population in the area where the land is best for farming. In the delta and the lower Ganges Valley, rice is the most important crop and provides the principal diet for the people

Name_______________
Date_______________

of India. In the eastern part of the delta, jute is the chief crop and most of it is made into burlap, a coarse cheap cloth often used to make gunny sacks. Jute is also woven into cloth for clothing and is used for twine, carpets, rugs, and linoleum.

India produces more tea for export (chiefly to Europe) than any other country in the world. The home market for tea is also important because it is the most popular drink in India. India is one of the leading producers of sugar cane which is grown on small plots in the Ganges Valley. Almost every village has a small sugar mill. Sugar from these mills is not refined. Instead, the stalks are rushed to squeeze out the juice which is then boiled down to make a soft brown sugar called *gur*. This is the kind of sugar that most of the people of India use. Although sugar cane production is large, India usually has to import sugar since most of the domestic production is used by the regions in which it is produced.

Modern industries are developing slowly. Today India produces textiles, steel, processed foods, cement, machinery, chemicals, fertilizers, consumer appliances, and automobiles.

Calcutta is India's largest city and leading seaport. It is built on the delta of the Ganges River, about eighty miles from the Bay of Bengal. It has rail connections with cities in all the important producing areas of India.

Bombay is India's most modern city and together with Calcutta handles four-fifths of all the foreign commerce of India. Some of the wealthiest people of India live in Bombay.

Railroads are an important form of transportation in India with 38,752 miles of track around the countryside. Passenger cars and other motor vehicles are in good supply.

India was under British rule from 1858 to 1947. After World War II, Indians worked to free themselves from British influence. Finally, through the efforts of Mohandas Gandhi and his nonviolent approach for obtaining India's independence, India became free. This was not without cost, however, because Gandhi was assassinated by a young Hindu fanatic in 1948. Jawaharlal Nehru became India's first prime minister.

Under Nehru plans were made to control the rapidly increasing population and establish a better agricultural system. A five year national development plan was set up mainly for irrigation and hydroelectric plants.

Nehru's daughter, Indira Gandhi, became the prime minister in 1966. During the 1970s, India's economy worsened. Many people were unemployed and food riots broke out. Suspicion grew against the government. Gandhi narrowly escaped losing her parliamentary seat. There was much unrest in India, particularly among the Sikh nationalists. On October 31, 1984, Indira Gandhi was shot and killed by a Sikh nationalist. Her son Rajiv Gandhi was sworn in as prime minister immediately after her death. Rajiv Gandhi was shot and killed by a Tamil suicide bomber. Narasimha Rao, a Gandhi supporter, became the prime minister. During the 1990s, India and Pakistan quarreled over who should control the Kashmir region of India. Atal Behari Vajpayee became prime minister in March, 1998. Today, India's government continues to have many problems to solve while building India's place in the world economy.

Name________________
Date________________

India

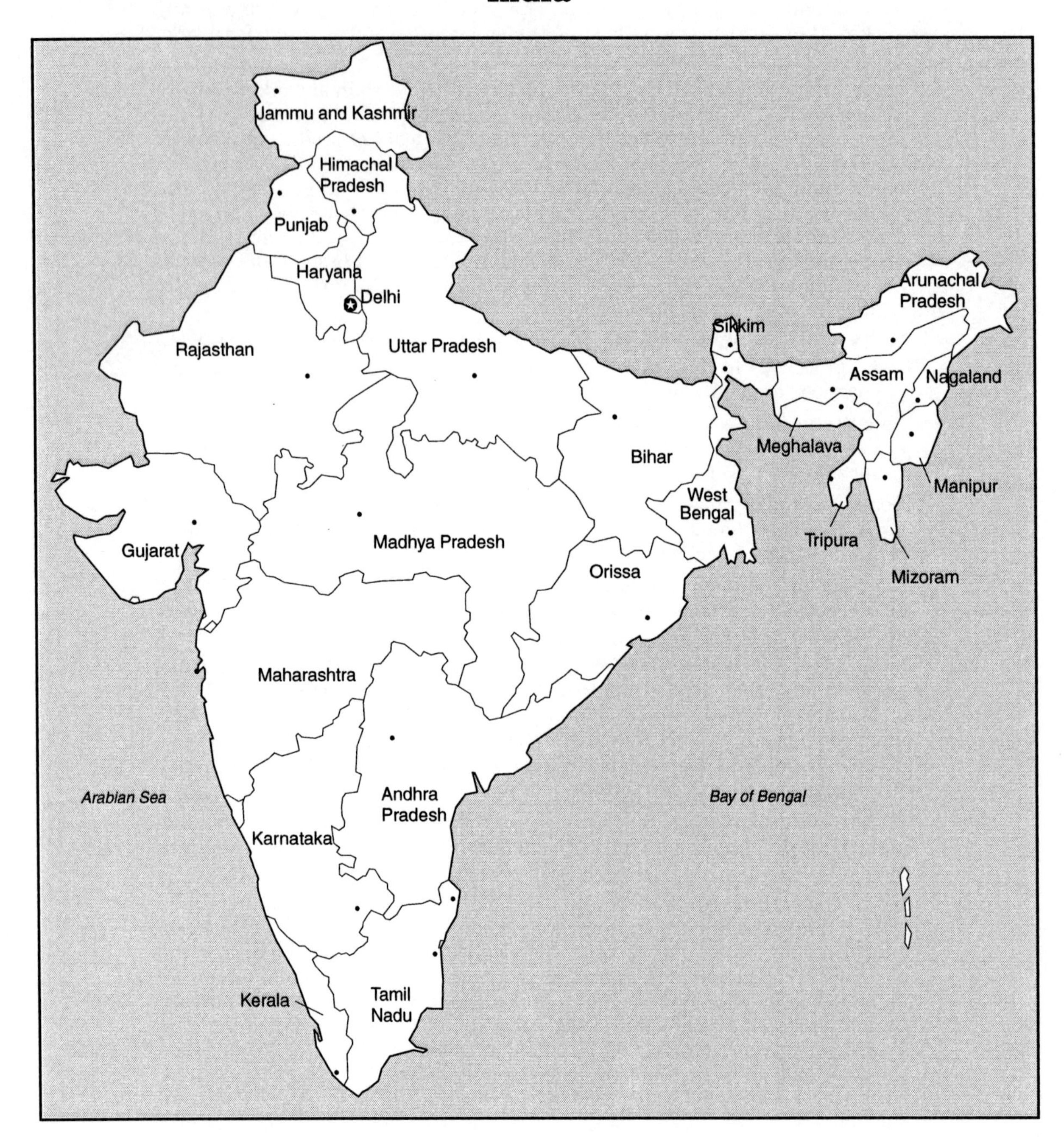

Name________________
Date________________

India—Creating a Panorama

Imagine that you are a geographer and your assignment is to illustrate the country of India so that visitors will be able to "see" the interesting sights and to read something about each one. Get ten sheets of heavy paper or poster board cut to 9" x 12" in size. Use the first sheet to make and illustrate a title/cover page. Include India in the title. On each of the remaining nine sheets make an illustration of one of the things listed below and research and write a brief paragraph about each one. Color the illustrations. Tape the ten pieces together as if folding a fan—one piece will go one way, the next will go the other way in a zig zag fashion—so that you have one long sheet of paper that will stand alone. (More paper can be used if you wish to make your panorama longer.) Select nine items from the list provided.

Sample

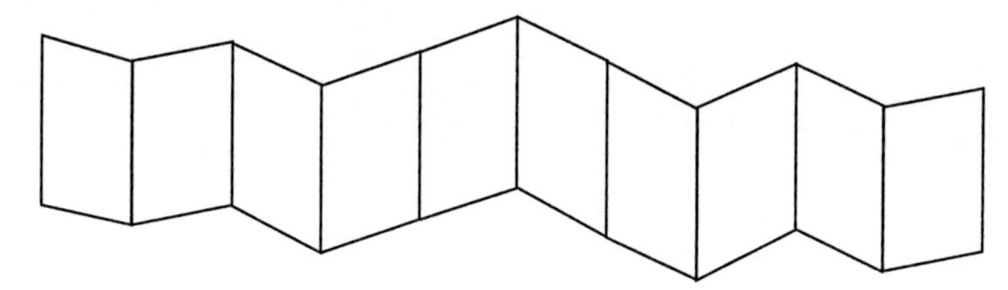

Places

In your paragraph, all places must have a latitude and longitude reading; population if it applies; climate; special features.

a. Himalaya Mountains

b. Ganges River

c. Calcutta

d. Bombay

e. Taj Mahal

f. Delhi

g. Great Indian Desert

h. Madras

Name________________
Date_________________

People

In your paragraph include geographic information such as birthplace, sites of important events, etc. It should be a brief biography, a good summary of the person's life.

a. Mohandas Gandhi

b. Jawaharlal Nehru

c. Indira Gandhi

d. Shah Jahan

Animals

Draw the animal in a natural Indian setting. Write about its adaptation to India's geography.

a. Tigers

b. Lions

c. Cobras

d. Mongoose

Lesson 26
Australia and New Zealand—Lands "Down Under"

Objectives

- To investigate major geographic, political, and economic features of Australia and New Zealand
- To compare the economies of Australia and New Zealand
- To examine the factors of production needed for the islands' industries

Notes to the Teacher

Australia and New Zealand are countries that generate a great deal of interest among students. In these "down under" lands, there are unique animals and seasons that are totally opposite those of the northern hemisphere. The language is full of unusual phrases that are distinctly different from other English-speaking countries. Australia, which is also a continent, and New Zealand have a wide variety of climatic and geographic differences that make them unique.

This lesson introduces students to the Pacific Rim countries of Australia and New Zealand. After reading some introductory material about the countries, students explore the countries' geography and economies to see how geography affects their resources and factors of production.

Students need atlases and some other research materials to complete this assignment. Students use a map of the two countries to locate major geographic, political, and economic features. Students read about the countries and learn some basic information. They complete a chart to answer questions about the two countries and their economies. Discussions continue throughout the lesson and students identify specific factors of production in the islands' industries.

Procedure

1. Display something that may have come from Australia or New Zealand. Kiwi fruit would be fun if available. Slice it and allow students to sample. Survey students to see if they know what it is and where it comes from. (New Zealand) Use discussion to find out what students know about this area. Display pictures of Australia and New Zealand showing major cities, rural areas, geographic features, etc. Have students guess how far away these countries are from students' home town. Then measure on a globe to get actual figures.
2. Distribute **Handout 68** to students. Have them label Australia's states, identify major rivers, mountains, lakes, major cities, and surrounding bodies of water. Use color. Indicate areas where products are produced—sheep, wheat, sugar, etc. Have them make a key to indicate use of symbols. Repeat this same procedure for New Zealand.
3. Have students read about Australia and New Zealand on **Handout 69**. Ask students to speculate why Australia and New Zealand are called the lands "down under." (*They are below the equator.*) Determine if any students have heard stories about these areas (*e.g.*, people walk upside down, etc.).
4. Discuss the reading and see what generalizations students can draw.

   Possible Generalizations:

   1. *Manufacturing is difficult to engage in when the population of a country is small (not enough workers).*
   2. *Mountains serve as barriers to keep people from settling in certain areas.*
   3. *People can adapt to a harsh environment in spite of the difficulties this presents,* e.g., *irrigation for dry areas.*

5. Provide students with **Handout 70**, and ask them to examine and use their own paper to answer the questions. They may work in small groups to complete the assignment. Debrief the class by reviewing completed charts. Use students' responses as a springboard for discussion.

Suggested Responses:

1. *Australia*
2. *Australia—more types of industries*
3. *Exporting goods brings money into the country's treasury.*
4. *Importing goods allows countries to get what they need when it is not available at home.*
5. *The disadvantages of importing goods is that it makes a country dependent upon another. In times of crisis these goods may become unavailable for importation.*

5. Tell students that they will need to identify the factors of production for some of the industries in Australia and New Zealand. Help them define the factors of production in this way:

   Land *natural resources (water, trees, soil, minerals, etc.)*

   Labor *human resources (mental and physical)*

   Capital goods *capital resources are human-made things like buildings, tools, equipment, and machinery*

   Production *combining productive resources to obtain goods and services*

   Have students look at the chart and note the industries. Working in groups, have students make three columns on their paper headed land, labor, and capital goods. Students should list under each heading what factors of production are necessary for each industry. For example:

   Dairy Products
   Land

   grazing land

   water

   Labor

   herders

   people to do milking

   people to drive tanker truck

   people to run machinery at dairy plant that prepares and packages the dairy products

   Capital goods

   tanker milk trucks

   milking machines

   dairy barn

   dairy plant

   machinery at plant

   packaging supplies

6. Ask students to share their answers about the factors of production. Conduct a discussion which helps students understand that each country has different resources. A country's economy depends upon these resources, and where they are in scarce supply a country must import those resources to meet their production needs. People often choose to live in specific areas because of the resources that are available to them. It is important to be able to participate in the trading process to acquire those scarce resources. This becomes a political issue based on interdependence in a global society.

Enrichment/Extension

1. Research and prepare a report on the following topics: Australia's sheep stations—
   - Discover their history and how they operate. Find out where their wool goes and current rates.
   - The production of wool and the process for producing it—What companies in the United States use wool?
2. Create a travel brochure for Australia or New Zealand. Make it colorful, interesting, and inviting to a traveler. Compute the actual cost from your home to either of these countries. Share your brochure with classmates.
3. Prepare a listing of some of the islands included in Australia's and New Zealand's territories. Select one and do further research. Prepare and deliver an illustrated oral report based on your findings.

Name__________________
Date__________________

Australia and New Zealand

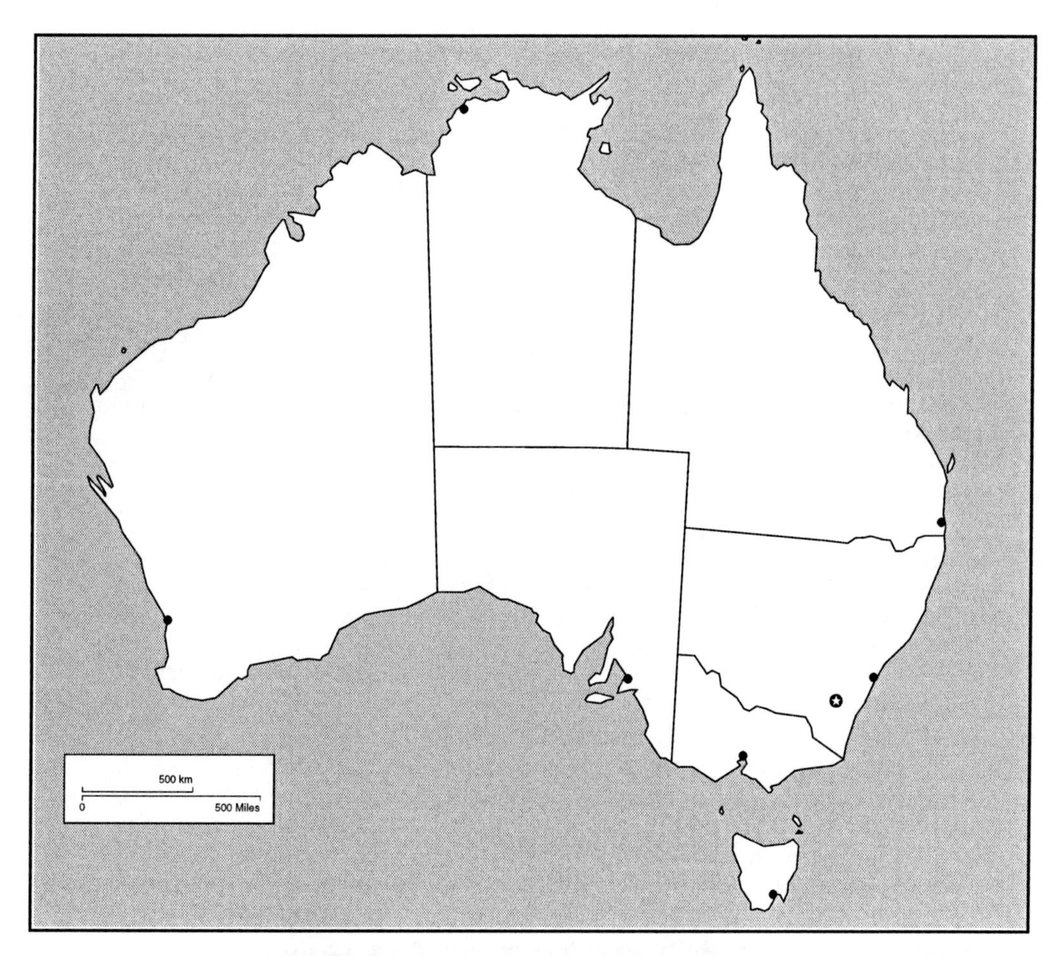

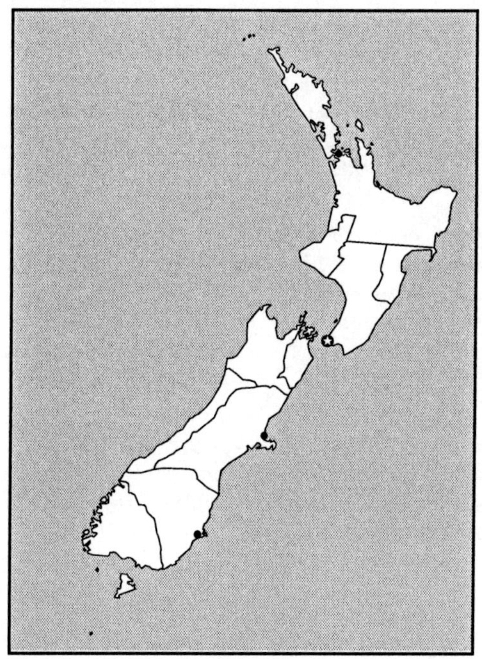

Name________________
Date________________

Exploring the Land "Down Under"

Read the following article and decide why these lands are referred to as being "down under." Also, think about how these lands are similar to many other places.

Can you name the continent that is located entirely south of the equator? If you said Australia, you were right. It is really an island, but it is so large, 2,966,200 square miles, that geographers classify it as a continent. It has a population of 18,077,000 people—many of whom are descended from people who came after Cook's voyages into the area in the late 1700s. Others are descended from some of its original settlers, the Aborigines, who came to the continent nearly 40,000 years ago.

Australia is the sixth largest country in the world (and the smallest continent) lying between 11 and 44 degrees south latitude. It is about the same size as the continental United States. Australia has six states: Queensland, New South Wales, Victoria, Tasmania (another island), South Australia, and Western Australia. It also has three territories on the main island and seven territories on other islands nearby.

There are mountains along Australia's east coast known to the settlers as the Great Dividing Range because they held settlers out of the land to the west. Beyond this mountain range are the central lowlands. Under this land can be found huge amounts of water, all underground. Farmers tap into it with windmills to irrigate their land and to find water for their animals. The rest of the continent, the very dry Great Western Plateau, is rich in minerals.

Australia's economy has always relied on its agricultural and mining products. In spite of the dryness of the country, raising sheep has been successful. Australia is number one in the world in the exportation of wool, with New Zealand right behind in second place. Australia also leads the world in the exporting of beef and veal, live goats and sheep, lead, and coal. Japan is its biggest customer and buys more than one fourth of these products. Australia imports heavy equipment from the United States and Japan.

Australia also exports wheat, beef, sugar, and dairy products because it produces more than its small population needs, and because the dry lands are farmed so extensively that great returns result. Sheep farms are known as stations. They raise about twelve sheep on every square mile of land. The stations may be tens of thousands of acres. Because the climate is unpredictable, the production of agricultural products is not constant, but varies from year to year. Thus, the money or income from the sale of agricultural products also varies.

New Zealand is an island country east of Australia. It is about the size of Colorado—14,454 square miles—with a population of 3,389,000. It is made up of two islands: North Island and South Island. Both islands are hilly and mountainous, but their east coasts are fertile plains areas. Like Australia, English is the principal language of New Zealand. The climate is generally mild with little seasonal differences.

New Zealand's most important resource is its land. It is suitable for farming and raising dairy cattle and sheep. New Zealand is the largest exporter of dairy goods in the world. In addition to wool, New Zealand exports kiwis, fish, lamb, mutton, and beef. It imports much of its manufactured goods, heavy machinery, and oil. New Zealand manufactures paper and paper products, lumber, meat, and dairy products, as well as other items, but there are not enough people or raw materials to support heavy industry.

Name________________
Date________________

New Zealand's capital, Wellington, is the shipping center. Another urban center, Auckland, serves as a port city and a dairy distribution center. Christchurch is the center for wheat and grain production while Dunedin is the wool and gold center.

Australia and New Zealand are considered to be part of the Pacific Rim region. Both countries have cultural roots in southeastern Asia and until their possession by Captain James Cook of Great Britain, no Europeans were interested in them. The areas have become highly "Europeanized."

The early settlers in Australia were farmers or gold miners and imported many manufactured articles. Some of the early settlers were also convicts from Great Britain. After the American Revolution, the British could no longer send their convicts to America. (Many reformers in the eighteenth century believed that the best way to prevent crime was to remove criminals from their country.) In 1786, Botany Bay in Australia was established by the British as a penal colony. There were many "criminals" whose only crime was being poor and in debt. Food strikes, harsh penal laws for minor offenses, and the fact that many people had lost their homes and employment due to the Industrial Revolution caused many families to be labeled criminals and to be deported to Australia.

Manufacturing developed in Australia, and today 36 percent of the labor force is involved in manufacturing compared with 20 percent in New Zealand. These two countries work largely in primary production: farming and mining today. They are getting in tune with areas of higher technology, and Australia is making more energy-producing goods such as coal, oil, and uranium. They have discovered a market for these valuable resources. New Zealand has recently discovered great natural-gas deposits on North Island but uranium sources believed to be present still remain undetected. Many changes lie ahead for these countries that lie "down under" from the United States.

Name________________
Date_________________

Australia and New Zealand—The Economy

Use the information on the chart below to compare the economies of Australia and New Zealand. Answer the questions that follow.

Type of Economy	Australia	New Zealand
Chief minerals	Coal, lead, iron, tin, copper, tungsten, zinc, uranium, petroleum	coal, limestone, gold, iron, petroleum, natural gas
Chief agricultural products	Wheat, meat, oats, butter, cheese, wool, livestock, fruits, hay, corn, sugar, barley, vegetables	meat, wool, butter, cheese, wheat, oats, barley, fruits, corn, vegetables
Industries and products	Meat, wool, dairy products, mining, food processing, iron and steel production, motor vehicles, heavy equipment, electrical appliances, aircraft, ships, textiles	frozen meat, dairy products, fishing, forestry, wood and wood products, textiles and clothing, transportation equipment, food processing
Chief exports	wool, cereals, meat, chemicals, hides and skins, ores and metals, fruits, vegetables, sugar, sugar products	meat, wool, butter, cheese, forest products, fish, kiwi fruit
Chief imports	transportation equipment, machinery, textiles, petroleum products	machinery, iron and steel, petroleum products, cars, plastics, chemicals, manufactured goods, textiles

1. Which country has the most goods to trade?

2. Which country appears to be more industrialized? Explain your answer.

3. How does exporting goods help a country? Explain.

4. How does importing goods help a country? Explain.

5. What disadvantages might be involved when a country must import goods?

6. What suggestions might you make to help Australia and New Zealand to improve their economies and make them more independent?

Lesson 27
Jordan—A Good Neighbor?

Objectives

- To examine the geographic region of Jordan
- To analyze the impact on human-environment interaction that takes place in Jordan

Notes to the Teacher

This lesson presents geographical background material about the country of Jordan. Consider adding some historical information and relating it, perhaps, to Jordan's historic role in the Bible—place names, in some cases, have been changed but others remain the same. Refresh students' memories about more current issues, such as the Persian Gulf War and its participants. The relationship that Jordan has had with the United States is another area with which students should be familiar.

Students read a short selection about the geography and the conflict in the region. There are a variety of facts presented and it should be pointed out how the geography affects the way people live and the decisions that they make about their role in the territory. Students are asked to imagine that they are editors of a local newspaper. They write an editorial about whether or not they believe Jordan is a good neighbor in the Middle East.

Procedure

1. Show students a shoe box full of sand. Place a tablespoon of black top soil or peat in one corner. Tell students that the box represents a country and that the black soil represents the only part of the country that one can farm. Make an indentation down the box about a third of the way in and trickle some water through here. Tell them that this is the major river in the country and the lifeblood of its agriculture. Ask students if they can guess in what part of the world this country might be located. Then, if they cannot guess, tell them that it is Jordan and explain how that conclusion was developed from the clues presented.

2. Use atlases and locate Jordan with absolute coordinates (31° N, 37° E). Then discuss the border countries. Ask students to relate what they already know about Jordan or any of its neighboring countries.

3. Distribute the reading, **Handout 71** and allow students time to read it. Then talk about the climate and geography.

   Ask students the following questions:

   What effect do climate and geography have on the human population? (*affects housing, way of life, politics, desire for control of land and water resources*)

   Why would the Jordan River be important to Jordanians? (*supplies water for irrigation*)

   What industries does Jordan have? (*oil refining, tanning*)

   What religion do most Jordanians practice? (*Islam: Sunni Muslim*)

   Who is the king of Jordan? (*Abdallah II*)

   Why do you think the previous king, King Hussein, tried not to take sides in the Persian Gulf War? (*He didn't want to offend Iraq, his close neighbor, for fear Iraq would retaliate on Jordan.*)

   What happened in Jordan when so many refugees came there during the Persian Gulf War? (*Many people were unemployed, the value of their money went down, and many economic problems resulted.*)

4. Discuss current events related to this region. Then, distribute **Handout 72**. Ask students to consider the discussion and the reading and write about whether or not Jordan could be considered a "good neighbor" by its Middle Eastern neighbors. Students must back up their decisions with explanation.

5. Share completed editorials with the class. Allow students to discuss and argue their points of view. Take a vote as to how many agree or disagree as to Jordan's status as a "good neighbor."

Enrichment/Extension

1. Research the Hashemite Kingdom of Jordan's history and make a timeline of major events.
2. Research the life of King Hussein and his family.
3. Make graphs that illustrate Jordan's population, industry, and agricultural products.
4. Investigate the lifestyle of the Bedouins in Jordan and their importance to the kingdom.

Name________________
Date________________

Jordan

Read the article below and answer the discussion questions assigned by the teacher.

Would Jordan be considered a good neighbor if you lived in one of the countries located on its border? Let's decide.

Jordan, with a land area of 34,352 square miles (just a little bigger than Indiana), sits on an arid plateau with Israel to the west, Saudi Arabia to the south, Iraq to the east, and Syria to the north. The Jordan River forms a boundary between Israel and Jordan and between Syria and Jordan in its northern section; the southern part of the river lies entirely in Jordan. The ownership of the Jordan River's Western Bank has long been disputed by Jordan and Israel. Israel has occupied the West Bank since the 1967 Six-Day War; Israel fought and won this war with Egypt, Jordan, and Syria.

Only four percent of Jordan's land is fit for farming, and most of that land is in the west. Its only seaport is on the Red Sea's Aqaba Gulf. A road linking the capital city of Amman to Aqaba in 1978 helped the port city's economy. Zarqa is Jordan's second largest city and where most of its industry has developed; it has an oil refinery and a tannery. The Dead Sea is shared by Jordan and Israel. The Jordan River flows into the Dead Sea.

Ninety-eight percent of Jordan's population is Arabic. Most of the population (92 percent) belongs to the Sunni-Muslim religion. Eight percent are Christian. Jordan's government is a constitutional monarchy headed by King Abdallah II. King Abdallah succeeded the previous king, Hussein, in February, 1999. He was born on November 14, 1935, and became king on May 2, 1953, when his father was declared mentally unfit to rule by Jordan's parliament. King Hussein was married to American-born Queen Noor.

The West Bank is an area of about 2270 square miles and has a population of nearly a million people, most of whom are Palestinians with a small Jewish population who have entered since 1967. (Before the British and the United Nations set up the country of Jordan and later Israel in the 1940s, all of the territory of these two countries belonged to Palestine. Palestine was dissolved, and Jordan and Israel were created out of the territory.) Jordan occupied the West Bank in 1949, but when they annexed it, most of the world would not recognize that it belonged to Jordan. In 1967, Israel captured it during the Six-Day War. In the 1970s, Hussein suggested a plan to unite the East and West Bank and call it the United Arab Kingdom. Other Arab countries, especially Egypt, Syria, and the Palestine Liberation Organization did not like the idea, believing that Hussein just wanted to control the land himself instead of letting the independent Palestinians control it. Jordan backed off in October 1974, and let the PLO take over the decision-making concerning who would control this area. By 1988, King Hussein gave up all claims on the West Bank. Israel and the PLO in 1993 signed an accord that gave the Palestinians self-rule in the Gaza Strip and the city of Jerico in the West Bank.

During the Persian Gulf War, which took place in 1991, following Iraq's invasion of Kuwait in August, 1990, the United States, Saudi Arabia, and other Arab countries were not happy with King Hussein because he tried to stay in the middle and not offend Iraq. Many Jordanians came back into Jordan during this period; they were fleeing from Kuwait and other Arab countries. Jordan was flooded with people. Their unemployment rate went up and the value of the dinar (Jordanian dollar) went down. Many economic problems came as a result. Jordan is still seeking solutions to these problems.

Name________________
Date________________

You're the Editor

Imagine that you are the editor of a newspaper in a country that borders Jordan. Write an editorial column concerning Jordan as your neighbor. Are they good neighbors? Explain exactly why you reached your conclusions.

THE NEWS GAZETTE

Editorial Page

Lesson 28
Egypt—A Strategic Connection

Objectives

- To examine the geography of Egypt and identify areas of human-environment interaction
- To demonstrate the value of a specific location
- To explore the importance of the geographic theme of movement

Notes to the Teacher

This lesson explores geographic information concerning the country of Egypt. It is important that students understand the geographic and historic background of Egypt. (In the Center for Learning publication, *World History and Geography, Book 1: The East*, lessons 4 and 5 are about hieroglyphics and the Suez Canal. These lessons may prove useful. Other lessons about famous Egyptians, Cleopatra and Sadat, can be found in the Center of Learning publication, *International Biographies: Africa and the Middle East.*) The Panama Canal lesson in this unit, lesson 12, might also be used in correlation to this lesson as a basis of comparison. Consider using Venn Diagrams of the two as a comparison tool.

Students read information about Egypt and the Suez Canal. Then a series of discussion questions are asked to determine students' understanding. A map activity follows in which students are asked to use their geography skills in relation to Egypt and the Suez Canal. The extension activities are helpful in students' total understanding of the region, so assigning at least one of these activities is recommended.

Procedure

1. Display pictures of Egyptian pyramids, the Nile River, the Aswan Dam, the Sahara Desert, Egyptian people, etc. Encourage students to discuss what they know about Egypt. Help them understand that Egypt's cities are modern even though 60 percent of the people are agriculturalists. Students often think that things have not changed very much from the days of the pyramid builders.
2. Ask students to read **Handout 73**. When they have finished reading, conduct a discussion around these questions:
   - What is the absolute location of Egypt? (25° N and 30° E)
   - What is the relative location of Egypt? (*Libya to the west, Sudan to the south, the Mediterranean Sea to the north, and Israel, the Gulf of Aqaba, and the Red Sea to the east*)
   - What problems has Egypt had with its neighbors? (*It often has territorial disputes with neighbors, especially Israel as Israel tries to add to its territory. There are also political difference among Egypt and the other Arab countries.*)
   - Is Egypt a good place to be a farmer? Why or why not? (*It depends upon where you are. The delta is great, but other areas must depend upon irrigation to water their fields. It is a hard life. There is not enough good land for all farmers to use.*)
   - Why are fewer than half of Egypt's population able to read? (*One reason is that school is only mandatory for a few years.*)
   - Why is the Suez Canal so important to Egypt? (*It is located on Egyptian territory, and Egypt can control the transportation that passes through it. This gives them great power over much of the world's economy.*)
   - Why does the world depend on the Suez Canal so greatly? (*The world depends on the Canal for movement—to have a shortcut to move heavy goods between the Pacific Ocean and the Mediterranean countries.*)
   - How do people depend on the geography and climate of Egypt? (*Once they depended on the flooding of the Nile for good soil and water; they live on oases where water is found for agriculture; they farm the delta's rich soil; they irrigate.; they keep animals and live nomadic lifestyles in the desert.*)
   - How have people changed the Nile so that the natural geographic features it once had are now under the control of humans? (*Aswan Dam*)

3. Using a blank outline map of the world, **Handout 74**, draw a line showing the shortest route for a Japanese tanker ship to come to a Mediterranean port to purchase oil. Label Japan, Egypt, Israel, Sudan, Libya, the Red Sea, the Suez Canal, Jordan, Saudi Arabia, the Mediterranean Sea, and the United States. Write a paragraph discussing the importance of the Suez Canal to the world economically and how this demonstrates the human-environment interaction, a geographic theme. Label the Panama Canal.

4. Have students read **Handout 75** in small groups; then have them develop answers to the following questions:

   - What changes might occur in the future that would affect the value of the Suez Canal? (*Cheaper air transport for heavy goods; new technology to transport oil without going through the canal; faster ships; goods supplied from other parts of the world, etc.*)

   - What would happen if one country took over the canal and charged very high rates for the right to pass through? (*Alternative methods of transportation would be needed causing the things in the first answer to perhaps happen even faster.*)

   - What consequences would there be if terrorists destroyed the Suez? (*Possibly there would be a war when the terrorists were identified; perhaps a new canal would be dug; or maybe people would just use other methods of transportation.*)

5. Have students explain in a paragraph how the Egyptians of today and yesterday depended on the geography and the climate of their country and how they interact with it to satisfy their needs such as the building of the Aswan Dam and the Suez Canal. Share with the class.

Enrichment/Extension

1. Research the countries that use the Suez Canal the most and discover what kinds of goods go through the canal.

2. Research Ferdinand de Lesseps and find out about his connection with the Panama Canal. (Lesson 12 in this book might be helpful as a follow up.) Give an oral report about the man and his canals. Which ones were successful?

3. Research the opera Aida and discover the connection between it and the Suez Canal in more detail. Retell the story of Aida and its first production. Consider doing a puppet show for the class.

4. Read biographies of presidents Sadat and Nasser.

5. Make posters using hieroglyphic symbols and pictures to illustrate something about Egypt—historic or modern.

Name________________
Date_________________

Egypt—A Strategic Connection

Why does this small nation hold power over so much of the world? What has happened to allow Egypt to have this control? Will this change?

Egypt is located in the northeastern corner of Africa. It is 386,650 square miles in area which makes it the twelfth largest country in Africa. Libya is located west of Egypt with Sudan to its south and Israel on its eastern border along with two bodies of water, the Gulf of Aqaba and the Red Sea. The Mediterranean Sea is its northern boundary.

Egypt is hot and very dry—mostly desert except along the Nile Valley where the land is green and fertile. The Nile River is the longest river in the world, and in the past it overflowed its banks during Africa's rainy season bringing fertile soil and water to a strip of land about six miles wide (narrower in some places) along the river. Today the Aswan High Dam controls the waters of the Nile, and flooding has stopped. The Delta of the Nile is the most fertile land found in Egypt. This fan-shaped piece of land is where 60 percent of Egypt's cultivation occurs.

The Sinai Peninsula is also part of Egypt but is separated from the main part of the country by the Suez Canal. It has three bodies of water on its borders: the Gulf of Aqaba in the east, the Gulf of Suez in the west, and the Mediterranean Sea in the north. Bedouins make up most of the population of this region.

Most of Egypt's population (90 percent) is of Hamitic Arab descent. Some Egyptians are of Nubian descent with dark skins and Arabic blood; others are Bedouin, a distinct, nomadic desert tribe. Camels are an important part of the Bedouin life. The principal language is Arabic and most Egyptians (94 percent) are Sunni Muslims. The population is growing very, very fast. The government (a republic) fears that food shortages will occur as a result. There is not enough good farmland available to grow food for Egypt's people. Only three percent of Egypt's land is arable. Therefore, Egypt already imports large amounts of food.

Children only have to attend school between the ages of six and twelve. Only 44 percent of the population is literate. A male's average life span is fifty-nine years and a woman's average life span is sixty-three years. There is only one telephone for every twenty-two people and only one television per eleven people.

Cotton is Egypt's major crop. Egypt also has textile, chemical, petrochemical, and food processing industries. Egypt produces oil, phosphates, iron, manganese, and limestone. Egyptians also raise cattle and have a fishing industry. Tourism plays an important part in the economy of Egypt.

Egypt and Israel were at war for almost thirty years, but a peace treaty was signed in 1979, and under Egyptian President Anwar Sadat's leadership, peace was established. The Sinai Peninsula was returned to Egypt by Israel. Over the years, Egypt has been a part of many battles over the control of its territory and that of the Suez Canal. In fact, on July 26, 1956, Egypt's Nationalization Policy enabled President Nasser to take control of the Suez Canal for the purpose of using the revenue gained from it to build the Aswan High Dam.

Israel invaded Egypt that October, and British and French military units attacked Egypt to regain the Suez. The Egyptians sank forty ships in the Canal to block it. The United Nations cleared away the debris and, in March of 1957, the Suez reopened. During the next year, the Egyptian Canal Company bought out the interests of the other countries in the Suez Canal,

Name________________
Date________________

and they were in charge of the whole operation. Egypt and Israel, however, continued to have conflict over the Canal. In 1979, Egypt finally agreed to give Israel unrestricted use of the canal.

Control of the Suez Canal gives Egypt much power over the countries surrounding the Mediterranean Sea. If they wish to move quickly from the Mediterranean Sea through the Red Sea to the Indian and Pacific Oceans, they must use the Suez Canal. Other countries also use this shortcut to enter the Mediterranean Sea area to trade. It is an important economic link to the rest of the world. The Canal has been enlarged and widened several times to allow larger boats to enter the canal. They intend to keep it in good repair and usable for modern boats. Thus, the Suez Canal will continue to play an important role in the world scene.

Name________________
Date________________

A Geographic View of the Suez

On the world map below, locate Egypt, Sudan, Libya, Israel, Jordan, the Gulf of Aqaba, the Red Sea, the Suez Canal, the Mediterranean Sea, Saudi Arabia, Japan, and the United States. Draw a line from Japan to show the shortest way for a Japanese oil tanker to travel to a Mediterranean seaport to buy oil. Locate the Panama Canal. How does this canal help ships?

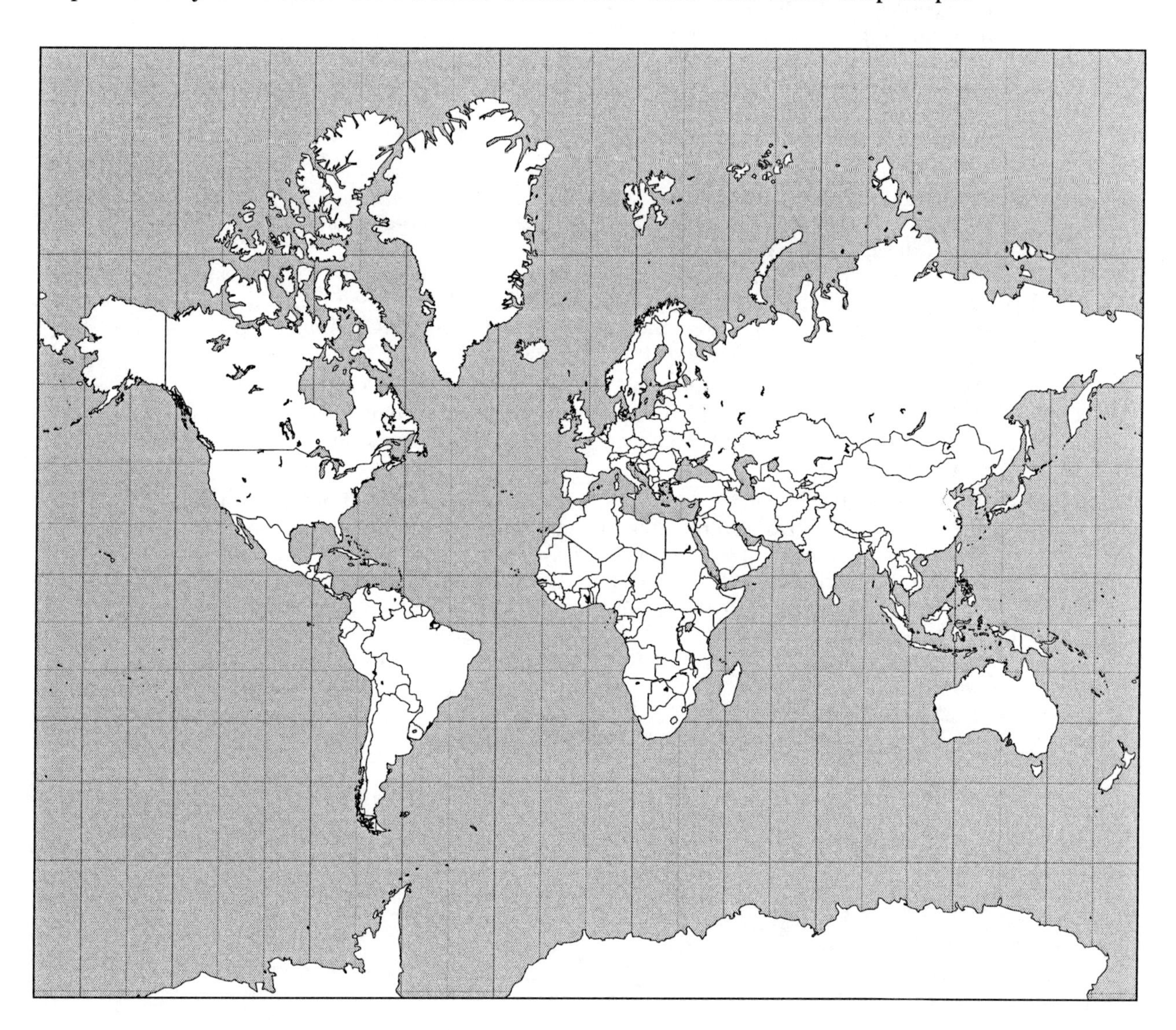

Write a paragraph explaining how important the Suez Canal is to the world economically. Explain how this canal is an example of human-environment interaction.

Name________________
Date________________

The Suez Canal

Read the following article with your group and answer the discussion questions assigned by the teacher.

The Suez Canal was the idea of a French diplomat and engineer, Ferdinand de Lesseps. In 1854 he talked the Egyptian Khedive, Said Pasha, into allowing him to start the project. The agreement allowed the company to own and operate the Canal for ninety-nine years with the Egyptian government regaining control at the end of that time. This private company was made up of Egyptian and French investors. (Financial problems caused the company to lose control, and in 1875 the British government bought up the stock and took control.)

Ferdinand de Lesseps set up the Universal Company of Maritime Suez Canal in 1858 and sold $40 million dollars worth of stock to build the canal. They began digging the canal in 1859 and took ten years to complete it. There are no locks in the Canal. It is built in three sections beginning at Port Said on the Mediterranean Sea and ending at Suez on the Red Sea in the south. It is over 100 miles long and is more than 60 feet deep. Today ships weighing up to 150,000 tons can make the passage through the Canal.

The first fleet of ships sailed from Port Said on November 17, 1869, for the Red Sea. During the passage, the great Khedive gave a ball to entertain the passengers. The opera Aida was also written and performed for the opening of the Suez Canal.

At an international conference in 1888 it was agreed that *all* nations could use the Suez Canal at any time. Great Britain maintained the military defense of the canal zone. From 1948, when Israel was formed out of Palestinian territory, their vessels were not allowed in the canal. By 1956, all British troops had left the canal zone at the insistence of the nationalist forces of Egypt. After many years of conflict with Israel, Egypt signed a peace treaty with them in 1979, and the Israelis now have unlimited access to the Canal.

Lesson 29
Harambee—Kenya's Many Cultures

Objectives

- To identify Kenya's various ethnic groups and their geographic locations
- To examine and analyze the human-environment interaction and the movement of people and ideas

Notes to the Teacher

This lesson looks at Kenya's variety of geographic regions and the various ethnic groups that reside in Kenya. Review with students the period of colonial imperialism by European countries (*e.g.*, Britain, Germany, Belgium, and France) in Africa in the 1880s. Remind students how Africa was divided among European countries to attain the natural resources and to gain new trading partners. Help students understand how colonialism caused Africans to lose their freedom and their culture through the governmental powers of the colonials.

Students read a short selection which is a summary of information about Kenya, and study a chart of the ethnic groups that live in Kenya. Discussion of the reading and questions about the chart follow. Using atlases, students draw a map of Kenya and indicate the boundary countries and selected sites within Kenya.

Procedure

1. Display pictures of Kenya showing Mount Kenya, highland areas, grassland areas, Lake Victoria, various ethnic groups such as the Masia, Kikuyu, etc. Ask students to guess which country they would travel to in order to see such sights.
2. On a globe, locate Kenya. Discuss the location of the equator in central Kenya and the kind of climate one would expect to find. In students atlases, find a climate map of Kenya and examine the various types of climate.
3. Distribute **Handout 76**. After students have read the information, discuss the present status of Kenya and tell students about the colonial period that began in the 1880s. Explain that European countries wanted the resources they found in Africa and they wanted a place to sell goods as well. Be sure students know that it was not until the 1960s that most of the African countries got their freedom either by fighting or by exerting political influence to get independence and rule by the black majority. Be sure to note the meaning of "harambee!"
4. Give students a sheet of posterboard, 18" by 24". Have them draw a large map of Kenya and indicate its border countries/water boundaries. Then, locate and label the capital city and other large cities. Refer students to the physical map. Have them color with colored pencils the grasslands, the highlands, and other physical features. Have them make a key to indicate the color code. Consider doing this assignment in small groups.
5. Distribute **Handout 77** and have students answer the questions about the chart independently. Then in small groups, have students discuss their answers to see if they agree or disagree with each other. Each group should develop a consensus answer for each question. Debrief with the whole class. Use the climate maps and the ones students have made and discuss how humans can adapt to conditions that are not the easiest in which to survive as they locate where these groups live. Explore how people moving into these places affects the lives of other animals and plants in the area.

Suggested Responses:

1. *Kikuyu; They can get their candidates elected if they turn out to vote.*
2. *Pakot, Masai; Cattle, need only grassy areas; They can move the cattle around to find grass during dry periods.*
3. *Bantu*
4. *El Molo, Segeju; Wars have taken the lives of their young people and some have married into other groups.*

6. Conclude by having students write a paragraph speculating what they think Kenya might be like for these ethnic groups in fifty years. Will things change? Will it still be a peaceful unity among the groups? What could cause the situation to change? Will humans and animals be able to coexist without causing more animals to become extinct? Will "harambee" work forever? Provide class time so students can share and discuss their paragraphs.

Enrichment/Extension

1. Locate on your map of Kenya the national parks and reserves. You will have to research to find the sites.

   The game parks are Aberdare National Park, Amboseli National Park, Sibiloi National Park, Samburu National Park, Tsavo National Park, Lake Nakuru National Park, Masai Mara National Reserve, Meru National Park, Mount Elgon National Park, Nairobi National Park, and Saiwa Swamp National Park.

2. Research some of the ethnic groups in depth to discover more about their lifestyles, occupations, political views, etc. Present your findings to the class.

3. Make a mural showing the animals in Kenya who are in danger of extinction along with others who are not. Code the endangered species in a way that is obvious which are endangered. Research the reasons for their being endangered.

Name________________
Date________________

Kenya: A Lesson in Diversity

Read about Kenya. Be ready to discuss what you learn.

Kenya, an African country located along the eastern coast of this huge continent, shares borders with five other countries. Sudan and Ethiopia are north of Kenya; Uganda is west; Somalia is east; and Tanzania is due south. The Indian Ocean borders Kenya's southeastern coastline. The equator divides the country in half.

Kenya covers an area of 224,960 square miles—almost the size of Texas. It has four distinct regions—the coastal region in the southeast along the Indian Ocean, an interior plain with bush and grassland, a highland area in the southwest, and an arid plateau area in the north above the equator.

The Rift Valley in Kenya is one part of what is known as the Great Rift Valley that begins in southwest Asia and continues through east Africa as far as Mozambique. This great 4,000 mile long valley was created by volcanoes and has a variety of altitudes as it cuts its way through Africa.

The water system in Kenya drains into three different areas. The coastal waters flow into the Indian Ocean, the Rift Valley streams run into a system of small lakes, and in the lands west of the Rift Valley, water drains into the great Lake Victoria. The Tana River is Kenya's longest river and flows south and east where it fully empties into the Indian Ocean. Nairobi, its capital, was founded in 1899 when the British government built a railroad from the interior to the coast. It is one of the fastest growing cities in the world. It is located at a higher elevation, so the climate is pleasant with temperatures ranging from 57° to 70° F. Nairobi comes from the Masai word *Nyrobi* which means "the place of cool waters." It is where they brought their cattle to drink.

Many people from the rural areas have come to Nairobi looking for work in some of its industries. The rural poor have built shanty-towns on the outskirts of the city because there is a shortage of low-cost housing. Crime rates have risen, and many of the wealthier families have built walls around their homes and have hired armed guards to keep out thieves. Even so, life is better for these people because they earn more money than they would in the countryside. Government buildings are located in Nairobi along with many other modern buildings, an international airport, a university, and a top level museum. A special landmark is the Kenyatta Conference Center.

Kenya declared its independence from British colonial rule on December 12, 1963. Jomo Kenyatta, leader of the KANU (Kenya Africa National Union) political party representing the large ethnic groups of Kikuyu and Luo people, was released from his political prison to become their prime minister. Thus, the black majority would finally rule their country. Kenyatta changed the old colonial economic and cultural systems left by the British. Now Kenyans controlled the farms and the businesses formerly run by Europeans. Education was provided for everyone who wished to attend school, but school was not mandatory. (Today about one half of Kenya's population is literate.) Kenyatta was a powerful leader who ruled until 1978 when he died at age 86. The new president and the second person to hold this position, Daniel T. arap Moi, had served as Kenyatta's vice president and was elected with no opposition. Kenya is a one-party political system with some rivalry within that party. All Kenyans age eighteen and over may vote.

Kenyatta traveled around the country while he was president urging the many different ethnic groups to work together and to get along. He used the Kiswahili word *harambee* as the center of his message to unite. It means *pull together*. It worked. (Today *harambee* is the nation's motto.) Some believe it was through Kenyatta's personality that people have lived and worked

Name________________
Date________________

together since independence and these same people worry that President Moi will be unable to maintain this state of unity.

Kenya is a country with over thirty different ethnic groups and several languages. Kiswahili and English are the official languages of Kenya. Kiswahili[1] is the language used in business. It is a Bantu language that borrows many words from the Arabic and originated along the eastern coast where the Arabs traded with the Africans. English is the major language in the large cities.

Besides the Bantu speakers (58.5 percent of the population with the Kikuyu being the largest Bantu group: 21 percent of the population), there are the Nilotic and Cushitic speakers. The Nilotic people, coming from the Nile River area of Sudan, make up 11 percent of Kenya's population. The Luo group is the largest Nilotic group with 13 percent of the population. The much smaller Masai group belongs to the Nilotic group also and is probably the best known ethnic group. The Masai are very traditional, avoid the modern world, and have continued to herd cattle on the land set aside for them. The Cushitic people are the smallest group and are from two different areas: Ethiopia (the Oromo group) and Somalia (Somali). They are nomadic herders of sheep, goats, and camels. These three groups represent the three main language groups found in Kenya.

The population is growing in Kenya as medical and health facilities improve, and this fact is causing stress on the wildlife because people are moving into their territory. Many animals are at the point of extinction because they are being hunted either for food or as pest control. On the grasslands and bush country of Kenya there are many game parks and preserves where animals can find protection. Of course, there is also danger from poachers.

Kenya is a country with a long history. The oldest human fossils have been found in Kenya. Centuries of unwritten African history have been passed down in oral form—stories, folktales, music, etc. The modern boundaries of Kenya are new, but the country is old and has undergone many changes over time. What will Kenya be like a century from now?

[1]Kiswahili not Swahili—Many people speak about this language as Swahili but the word belongs to a group of nouns that always start with the prefix "ki." Technically Kiswahili is the correct form. For example, the language of the Kamba tribe is Kikamba and the Swahili tribal language is Kiswahili.

Name________________
Date________________

Ethnic Groups in Kenya

Using this chart and an atlas, answer the questions that follow.

Ethnic group	Language group	Location	Economic base	Special notes
Kikuyu	Bantu	Upper Lake Victoria, Central Kenya, Southern Coast	Farming	Largest groups of Bantu speakers; most political influence; 5,930,610 population
Segeju	Bantu	Southern Kenya	Cattle herding	War-like; Fewer than 1,000 left in Kenya
Taita and Taveta	Bantu	Taita Hills	Farming	Complex religion; talk to dead for advice; skilled basket weavers; approximate population 16,000
Tharaka	Bantu	Plains East of Mount Kenya	Beekeepers	Much malaria and sleeping sickness; known for witchcraft; approximate population 10,000
Boni	Cushitic	Interior coastal region	Farming, collect wild honey and fruit, hunt small game	Originally were hunters andgatherers; ivory traders; approximate population 5,000
El Molo	Cushitic	Islands in Lake Turkana	Fishing	May be descendants of Kenya's original inhabitants; only a few hundred left
Ilchamus	Nilotic	Shores of Lake Baringo	Farming, fishing	Noted for their dances; approximate population 7,000
Nandi	Nilotic	Nandi Hills	Farming	Were fierce warriors
Pokot	Nilotic	Great Rift Valley	Hill areas: farming; Plains: cattle herding	Raid cattle of Turkana tribe when searching for water and grass; approximate population 200,000

Name________________
Date________________

Ethnic group	Language group	Location	Economic base	Special notes
Samburu	Nilotic	Highlands around Mount Nyiru and arid plain below	Farming	Look like Masai in appearance, but more peaceful; approximate population 70,000
Turkana	Nilotic	Along shores of Lake Turkana	Cattle and camel herding/fishing	Very primitive group; live in remote area; approximate population 300,000
Masai	Nilotic	Reserved lands in southern Kenya	Cattle herding	Tall, slender, handsome people; traditional lifestyle; once were fierce warriors; less than 250,000 population

1. Kenya's current population is 30,765,916. Which ethnic group makes up the largest percentage of that population? With such large numbers, how does this explain their influence in a republican form of government?

2. Which of these groups live in the *arid* three-fifths of Kenya? (Hint: Use the climate map in atlas.) How does each group make its living? Why do these chosen occupations make sense (or not make sense) in these regions?

3. Which language group has the most people belonging to it?

4. Which groups are shrinking in number? What do you think is happening to them?

5. Pick one group. Write a law that they might like to see enacted. Then, explain why this law might be needed.

*Chart adapted from Robert Pateman, *Cultures of the World—Kenya* (New York: Marshall Cavendish, 1993), 64–65.

Lesson 30
Zimbabwe—Changing Times

Objectives

- To examine the geographic, historical, cultural, and economic features of Zimbabwe
- To make economic decisions about the use of natural resources

Notes to the Teacher

In this lesson students learn about the country of Zimbabwe—its geography, a little about its history and the people, and how they interact to affect Zimbabwe's economy. A review of some basic economic concepts such as the need for natural resources in the manufacturing process, the need for a labor force, and the importance of fertile land for agricultural pursuits is useful. Additional information about the history of Zimbabwe and its struggle for independence and black majority rule would be helpful to stimulate students' thinking.

Students read a handout with basic information about Zimbabwe which can be used as a basis for discussion and the beginning of further research if that is desired. Students work in groups and use critical thinking skills to do problem solving in a simulation activity.

Procedure

1. Use a world map or atlases to locate Zimbabwe. Ask students to give its absolute location. Have some pictures of Zimbabwe that include the High Veld, the cities and rural areas, mining sites, etc. Old *National Geographic* magazines are a good resource.
2. Ask students if they think Africa's political boundaries always looked as they do now. (Most will respond that the boundaries have changed.) If possible show an old map of Colonial Africa. Have students tell why they think the boundaries changed. Discuss why some nations want to take over other nations.
3. Have students read **Handout 78**. Allow time for discussion and thoughtful questions. (For example, why would the British take the land from the local people and give it to the Europeans? Why was a white majority government an unfair way to run Zimbabwe? Why was the rest of the world so against this form of government?) Talk about the climate and the geographic features. Discuss what positive economic factors are available in Zimbabwe (arable land, rich mineral resources, good growing climate, large labor supply, etc.)
4. Have students do library research on Zimbabwe related to the simulation on **Handout 79**. Give groups enough time to discuss the problem and to research and report on their conclusions. Be sure one member of each group serves as a recorder, one as a reporter, one as the organizer of the research team and the information that is collected, and one as the member responsible for keeping everyone on task.
5. Have groups present their decisions/arguments to the class. Allow time for class questions. Teams should be ready to defend their positions.

Enrichment/Extension

1. Research and prepare a biography of Cecil Rhodes and present it to the class.
2. Research the two major ethnic groups in Zimbabwe: the Shonas and Ndebeles. Report on their culture and their political views.
3. Report on the mining of gold and platinum in Zimbabwe. Be sure to emphasize the economic impact of mining on Zimbabwe's economy.

Name_______________
Date_______________

Zimbabwe

Read the article below and be ready for a discussion.

Zimbabwe is a land-locked country in the southern part of Africa. Its northern boundary is the Zambezi River which separates Zimbabwe from Zambia. The Limpopo River is its southern boundary and it separates Zimbabwe from South Africa and Botswana. To the east and north is Mozambique with highlands along the border. Zimbabwe is part of the large plateau that makes up the southern part of Africa.

Most of Zimbabwe is tropical grassland. Forests are found in the Eastern Highlands and in the west there is a savannah woodland. In the middle of Zimbabwe is an area known as the High Veld, a giant hump of land that extends about 400 miles in a north-south direction. It is about fifty miles wide. The climate is perfect for European-style farming. On either side of this hump is the Middle Veld, a grassland at a lower latitude. Below this is the Low Veld. It is at a low altitude, has a hot, dry climate and very poor farming conditions.

Zimbabwe is an old culture that extends back into the Stone Age. In fact, its name comes from the original settlers in the area that built a huge stone fortress and houses that was known as The Great Zimbabwe, about 800 A.D. Most of the people who live in Zimbabwe come from the Shona people whose word Zimbabwe means "royal court." They make up about 80 percent of the population today with the Ndebele people making up about 19 percent and the European population stands at about 1 percent. The Ndebeles were a Zulu clan who fled South Africa when Shaka Zulu's army grew powerful enough in the early nineteenth century to take over the group. Land shortages were also a problem.

Europeans were interested by the late 1800s in taking over this area to get the gold resources and other minerals that were to be found. Cecil Rhodes from Great Britain sent prospectors and a private army into Zimbabwe to take over the land. The Shonas allowed the British in their part of the territory, and their agreement kept the British away from the Ndebeles' territory.

Many prospectors were disappointed that the gold mines were often mined out. Then people began to realize that this was a good place to farm. After some fighting between the British, the Shonas, and the Ndebeles, the British finally gained total control, renamed the country Rhodesia, and took the best lands to give to the Europeans to farm.

In 1965, the Rhodesians declared their independence from Britain because they feared that the British would allow the black majority to take over the country. When the white minority took charge, the world's opinion of Rhodesia dropped. The United Nations asked for sanction—no one should do business with Rhodesia—but sanctions were not enforced very well, and Rhodesia was not very affected. After much bloodshed, the white minority and the black nationalists finally sat down in 1979 in London and worked out an agreement that allowed the black majority to take over the government of their country. Independence came on April 18, 1980. Today its government is a parliamentary democracy with President Robert Gabriel Mugabe as its president. Its capital, Harare, has a population of 1.2 million.

Name________________
Date________________

The government is still deciding how to return some of the land to the poor rural farmers whose land a century ago was taken by the whites. The whites were given ten years from 1980 to keep their land, but in the last several years some of "their" land has been returned to the black majority. It is not known how much more land will be redistributed, but many of the nationalists are anxious to see this happen because land was one of their major reasons for seeking a black majority rule.

The country changed its name to Zimbabwe from Rhodesia to indicate that the British colonial rule had ended. Today English is still one of the official languages, however, along with the Shona and Ndebele languages.

The future of Zimbabwe weighs heavily on the successes that they have experienced since their civil war. They have tried to rebuild their country by making their farms more prosperous—they even export some food. The old villages have more food than they need for a change. During the embargo, they started many factories. The goods they produce are shipped to nearby countries. Their major problems are trying to maintain good relations between the Shona and Ndebele people and between the black Africans and the white minority group. The outlook is good for Zimbabwe. They are setting an example for other African countries, especially South Africa where racial problems are the major issue.

Name________________
Date________________

Zimbabwe
A Conference Simulation—A National Council Meeting

You are to read the role card assigned to your group. Discuss the details on the card with your group. Visit the library to read more about Zimbabwe before making your decisions. As a group, explain your decision to the whole class as if you were presenting it to Zimbabwe's National Council.

Ivory Dealers

Your business has dropped off over the last five years. Why? In 1989 there was an international agreement that banned the trade and sale of African elephant ivory. Animal rights groups caused the ban with their protests against the killing of elephants just to get their ivory tusks. Half of Africa's elephants have disappeared over the last ten years. As a country, Zimbabwe has been against this ban. Zimbabwe believes that each country should be looked at individually. They have a large game preserve where there are many elephants. Each year they must cull the herd (kill some of the elephants to reduce the number in the herd). If they do not cull the herd, the elephants will starve because of an inadequate food supply for a large number of animals. Ivory harvested from the dead elephants still could be sold if not for the ban. The price of ivory has dropped greatly and countries that used to buy it (China, Hong Kong, Japan, United States) are no longer interested in it.

Your Problem—What arguments could you present to Council that would convince them to allow you to buy and sell African ivory items such as jewelry, chess sets, statues, etc.? How would you get the international ban lifted? How would you advertise to let people know that the ivory was not taken by poachers?

Land Resettlement

You represent poor, rural farmers whose family's land was taken by the Europeans over a century ago and who were forced to farm on a "reserve" in a remote area where the land was poor and unsuitable for farming. Now that there is black majority rule in Zimbabwe, you want better land. Some people have been moved in one of three government resettlement plans: (a) some are placed in a village with individual pieces of land for crops or grazing, (b) some are placed in communal living situations and farm cooperatively, and (c) some are given their own land but it is part of a large estate that is run on a cooperative basis. Remember that the Europeans still hold most of the best farmland.

Your Problem—Things are moving too slowly. You want your own land as soon as possible. What can be done to make this change more quickly? How can this be done without causing trouble between the Europeans and the Africans? What will you say to the Council?

Name________________
Date________________

Growing Tobacco—A Cancer-Causing Agent

You represent the tobacco growers in Zimbabwe. You have heard that there may be an international ban on tobacco sales because of its link to cancer. Pressure is being placed on Zimbabwe to stop exporting tobacco. More than a million people work on the tobacco farms in Zimbabwe. It is one of the country's most valuable cash crops, and it brings in foreign currency when it is exported—making it important to the government. This tobacco is of very high quality. European countries may even stop all forms of advertising for cigarettes.

Your Problem—What arguments can you give to the Council to continue the production of tobacco? What can the government tell the European countries about the use of tobacco and cigarettes? How can you get around a ban on cigarette advertising in Europe?

Mining—What to Mine?

You represent asbestos miners. Over the last ten years, people in the United States and Europe have been concerned about the effect asbestos has on the health and environment of their people. In fact, the price has dropped because of these concerns. So far it has not been a disaster for miners because Zimbabwe still produces almost 200,000 tons of asbestos each year. There are many other kinds of mines in Zimbabwe: gold, coal, iron, nickel, platinum, and others.

Your Problem—You must convince Zimbabwe's Council that you must be allowed to continue to mine asbestos for export. Should your miners try to find work in other kinds of mines? Should they continue to produce asbestos? What can be done to convince people that asbestos is safe for use? Explain how asbestos is used. What could be done to make it safer?

Part 3
Hands across Our World

Suppose there were no nations, no borders, no barriers, no ideologies, no separation of the family of humanity. What would remain? People!

People the world over have certain universal cultural traits. They work, they play, they eat, they build dwellings, they worship a greater being, they produce arts and crafts, they communicate, they define their social roles. Each nation and culture generates its own variation on these cultural traits, yet they remain as the universal cultural traits of all humans. These cultural universals are what bind us together in the global village.

Lesson 31
Folktales—Stories that Teach

Objectives

- To identify the elements in folktales that are explanations of natural, geographic, or scientific phenomenon
- To develop insight into other cultures by examining their folktales
- To encourage creative thinking and writing skills

Notes to the Teacher

In this lesson students examine the culture of various groups through their folktales. They need to understand that folktales developed in cultures where written language had not yet become accessible to most of the people. These stories were used to pass down information to future generations about how their traditions and history came to be a certain way. The stories also offered explanations for how things happened that, without the scientific knowledge that we have today, they had no answer for, such as why crows were black and why mosquitoes bite.

It is important to provide books of folktales or time in the library to gather folktales to add to this lesson. Students are generally very interested in these stories. Many books on folktales can be found in the children's section of most libraries. Picture books are useful for this activity because the story is kept simple and illustrated to develop the concepts visually. Older students enjoy them also.

Students read a small sampling of folktales and discuss the elements that are meant to inform them. They must identify whether the information is cultural, geographic, scientific, or something else. Students create artwork that is related to the stories and a bulletin board with a world map showing where the various countries of origin are located. In addition, students will find a folktale on their own to present to the class and explain the information elements contained within it. Finally, students write their own folktale based on some modern tradition or phenomenon that they want the next generation to understand.

Procedure

1. Show students a picture of a plate with the blue willow pattern (or bring in a plate with the blue willow pattern). Let them examine it carefully noting the bridge, turtle doves, trees, etc. Ask students if they know the story of the blue willow pattern. (Most will not know the story.) Then tell them this story:

   In Old China the mandarins were very rich and powerful. One such mandarin had a beautiful daughter called Li-chi. The mandarin worked all day each day in his home and his secretary, Chang, a handsome young man, assisted him. Chang spent most of his free time on the bridge in the mandarin's beautiful garden. He liked to look at the water and watch the goldfish swim. Li-chi liked to stand on the bridge, too, but she liked to be near young Chang.

   They fell in love, but Chang knew that he could not marry the rich mandarin's daughter because he was just a poor man, but she said he has much more to offer than just money. Unfortunately, the mandarin overheard them talking about their love from his upstairs window. He was very angry and swore that no such marriage would ever take place. He made Chang leave his house. Li-chi cried sad tears.

   Chang returned that night in the darkness and begged Li-chi to run away and marry him in his home far away. She climbed down to join him. They hid in a dirty old gardening shack because Li-chi believed no one would look for them there. They waited all the next day while they listened to the servants search for them. When everything was quiet, they slipped out of the shack and started to cross their beautiful bridge from the island. Suddenly, the mandarin appeared on the opposite end carrying a huge whip. He shouted that he had trapped them. Knowing that they had no other way to get off the island and if Chang were to be caught he would be beaten to death, they jumped into the lake together. If they could not be together then they would die together.

The gods on the mountains saw what had happened and they loved them for their faithfulness and courage. Just as the mandarin's whip lashed out, Li-chi's arm turned into feather and Chang's body was covered in down. They were transformed into turtle doves and flew far away out of danger from the mandarin.

China's potters painted this wonderful but sad love story in blue on their best porcelain dishes. They sold these beautiful dishes all over China and other distant places.

2. Ask students what information they gleaned from this story. (*How the blue willow pattern originated among the Chinese.*) Why would people choose to give information in story form? (*Written language was not always available to the poor.*) Tell students that folktales were told because most people could not read or write in early times. Inform students that this kind of story has information about the culture of the people—how they lived. Other folktales give information about geographic, scientific, or other topics that people did not have the technical knowledge for so they used interesting stories to explain the phenomenon.

3. Have students read the Liberian folktale on **Handout 80**.

   Suggested Responses:

   1. *Women took care of the children; women looked after their husbands by preparing their food; women looked after the farms, raised the breadstuffs and fished, while men were supposed to hunt and bring meat for the family; people believed in supernatural powers to transform themselves; information given about keeping away from a leopard—climbing a tree.*

   2. *Cultural*

   3. *Women today share many jobs with men such as caring for homes and children; women work at careers outside the home; women have much more freedom to do as they choose.*

4. Have students illustrate the story and write a short summary of the story to go with the illustrations. Share and display their work.

5. Send students to the library to find a folktale. It can be a picture book or one found in an anthology without illustrations. If desired, students can be assigned a specific country from which the folktale must come. Be sure students read the folktale carefully and know it well enough to retell it and to discuss what it is trying to teach.

6. Indicate on a bulletin board with a large world map the country of origin for each story to be shared. Use flags or pushpins to note locations. Have students give locations in terms of latitude and longitude.

7. Share the folktales briefly with the class, making sure everyone understands the message being sent in the story. Invite students to think about some of the things in their culture, environment, geographical features of their area, scientific information, etc., that they believe should be passed to the next generation.

8. Have students brainstorm possible modern folktales topics. Write their suggestions on the chalkboard. Then have students create their own folktales and illustrate them. Pass completed folktales around for the class to read and enjoy.

Enrichment/Extension

1. Create a pattern for modern dishes with either a folktale you have read or the one you have written. Display the pattern and written explanations.

2. Examine myths from various countries and compare them to the folktales you have read. Select one folktale to retell and explain to classmates.

3. Examine a tall tale or legend from the United States. Retell one and explain how it is different from folktales of other countries.

Name________________
Date________________

The Leopard Woman
A Liberian Folktale

Read this Liberian folktale and answer the questions which follow.

A man and a woman were once making a hard journey through the bush. The woman had her baby strapped upon her back as she walked along the rough path overgrown with vines and shrubbery. They had nothing to eat with them, and as they traveled on they became very hungry.

Suddenly, emerging from the heavily wooded forest into a grassy plain, they came upon a herd of bush cows grazing quietly.

The man said to the woman, "You have the power of transforming yourself into whatever you like: change now to a leopard and capture one of the bush cows, that I may have something to eat and not perish." The women looked at the man significantly and said, "Do you really mean what you ask, or are you joking?" "I mean it," said the man, for he was very hungry.

The woman untied the baby from her back, and put it upon the ground. Hair began growing upon her neck and body. She dropped her loincloth; a change came over her face. Her hands and feet turned into claws. And in a few moments, a wild leopard was standing before the man, staring at him with fiery eyes. The poor man was frightened nearly to death and clambered up a tree for protection. When he was nearly to the top, he saw that the poor little baby was almost within the leopard's jaws, but he was so afraid that he couldn't make himself come down to rescue it.

When the leopard saw that she already had the man good and frightened and full of terror, she ran away to the flock of cattle to do for him as he had asked her to. Capturing a large young heifer, she dragged it back to the foot of the tree. The man, who was still as far up in its top as could go, cried out and piteously begged the leopard to transform herself back into a woman.

Slowly, the hair receded, and the claws disappeared, until finally, the woman stood before the man once more. But so frightened was he still, that he could not come down until he saw her take up her clothes and tie her baby to her back. Then she said to him, "Never ask a woman to do a man's work again."

Women must care for the farms, raise breadstuffs, fish, etc, but it is man's work to do the hunting and bring in the meat for the family.

1. List all the information this story is giving you about this group.

2. Tell in which category this information belongs: cultural, geographic, scientific, or another.

3. Compare/contrast this situation with women's roles in our society today.

Lesson 32
Languages—One or Many?

Objective

- To consider whether or not the United States should become a bilingual nation

Notes to the Teacher

When multitudes of immigrants in the nineteenth century were welcomed by Lady Liberty in New York harbor, they chose to come to the United States. These immigrants single-mindedly wanted to be Americanized. At the same time, they preserved their traditions and their native languages. Traditions remained, but younger generations who saw ethnic languages as a barrier against upward social and economic mobility eventually replaced ethnic languages with American English.

Today, however, pressure for bilingualism mounts. A new wave of immigrants from Spanish-speaking countries cling to Spanish. In New York, signs on shops, restaurants, banks, and churches are multilingual with languages of new wave Asian immigrants mingling with the rest. In the northeast, French is a second language in the industrial areas. In Florida, Spanish dominates; in Alaska, Eskimo, Indian, and Aleut are spoken by twenty percent of the people while Hawaii maintains English with multi-national variations. On the west coast, bilingual pressure was so strong that California enacted a law designating English, not Spanish, as the official language. By 1990, sixteen other states had passed similar laws or constitutional amendments. Whether or not this is a wise policy remains an issue.

For some, the choice seems clear-cut. Across the globe, now so closely linked by satellite television and rapid transportation, English has become the major international language. Diplomats to the United Nations speak English even though they use native languages in public discussion. Airline pilots and air traffic controllers communicate vital signals for landing and take-off in English. Latin American and Scandinavian countries require English as a second language. What, then, would be wrong with expecting newcomers to conduct their business, vote, and secure health care in the language of their new country?

Others, however, question the wisdom of this "English only" approach. Would it bring us closer together as a nation, or would it create further dissension? Does it suggest disrespect for other countries? Is it in our national interest to stamp out facility in other languages?

Students create a map of ethnic diversity in the U.S. and then brainstorm advantages and disadvantages of the United States becoming a bilingual nation. To conclude the lesson, students study the example of Australia, also a multi-ethnic nation, to see what lessons it can offer the United States in deciding the question of whether or not to become a bilingual nation.

Procedure

1. Ask if anyone speaks another language at home and what it is. Ask if students who do not have a second language in their homes would like to be able to speak one. Which one and why? (Example: Science and math students may need to know German and Russian to be able to do research in their areas.) Share the information in the Notes to the Teacher.
2. Distribute **Handout 81**. Tell students to color the various states according to minority languages spoken as indicated in the directions.
3. As a class, brainstorm advantages and disadvantages of having the United States become a bilingual nation. Put suggestions on the chalkboard as students generate ideas.
4. Distribute **Handout 82**. Have students read the background in preparation for a discussion on the following issues:
    a. How does the Australia policy indicate respect for different cultures? (*Willingness to promote and learn the language implies a respect for the culture and its potential benefits to society.*)

b. The elementary schools in Australia give special emphasis to nine "languages of wider significance": Arabic, Chinese, French, German, Greek, Indonesian/ Malay, Italian, Japanese, and Spanish. How might widespread knowledge of these languages promote the country's best interest? (*Knowledge of these languages could have political and business as well as social benefits.*)

c. Why would the country seek to preserve native languages of the Aborigines? (*Preservation of the language offers possibilities for studying ancient cultures and promoting good relations with the Aborigines. Australia may well succeed in averting the sort of disastrous relationship between its majority and minority cultures that has marred U.S. relations with its Native American population.*)

d. Does the Australian policy view languages as a problem or a resource? Explain. (*Clearly, the Australian government views language as a resource to be preserved rather than a problem to be eradicated.*)

e. How would you feel about the language policy if you were a new immigrant? Why? If you were one of the Aborigines? If you were an English-speaking Australian?

f. How might the policy be costly? Would the effort to promote diversity be worthwhile? Why, or why not? (*The teaching of many languages is costly but may result in long-term savings by promoting improved relationships between majority and minority populations and between natives and newcomers to the country.*)

5. To conclude the lesson, discuss whether or not the United States should become bilingual or adopt an English-only policy. Is there consensus within the class?

   Answers would include the following: Many Americans believe that recent immigrants should learn the language of this country if they expect to reap the benefits of life in America; they point out the inconvenience, cost, and divisiveness of a bilingual policy. Others point out that the America of today is vastly different from what it was in the past and that respect for the increasing Hispanic population, particularly in some parts of the country, dictates that the country rethink the long-established English-only policy and its political, economic, and social implications.

Enrichment/Extension

1. Do an investigative report on the number of different languages or ethnic backgrounds of your school. Your report could include a chart which would show the different languages spoken at home, the countries represented, and the number of students who can speak their ethnic languages fluently.

2. Poll your family and friends to help you compile a list of occupations in which bilingualism is a requirement. Have your classmates do the same thing. Make a composite list of these occupations. Write an essay in which you explain the implications for yourself and fellow students in preparing for jobs in the future.

Name_______________
Date_______________

Language Diversity in the United States

Color the map below according to the directions that follow.

- Using a **red** crayon or marker, color the following states where Spanish languages are spoken by a minority: one-half California, one-half New Mexico, Arizona, Texas, Florida, and Colorado.
- Using a **blue** crayon or marker, color the following states where French is spoken by a large minority of people: Maine, one-half Massachusetts, one-half Rhode Island, Vermont, New Hampshire, and Louisiana.
- Using a **yellow** crayon or marker, color the following states where Asian languages are spoken: one-half Massachusetts, one-half Rhode Island, New York, one-half California, and one-half Hawaii.
- Using a **green** crayon or marker, color the following states where Native American languages are spoken: one-half Hawaii, one-half New Mexico, and Alaska.

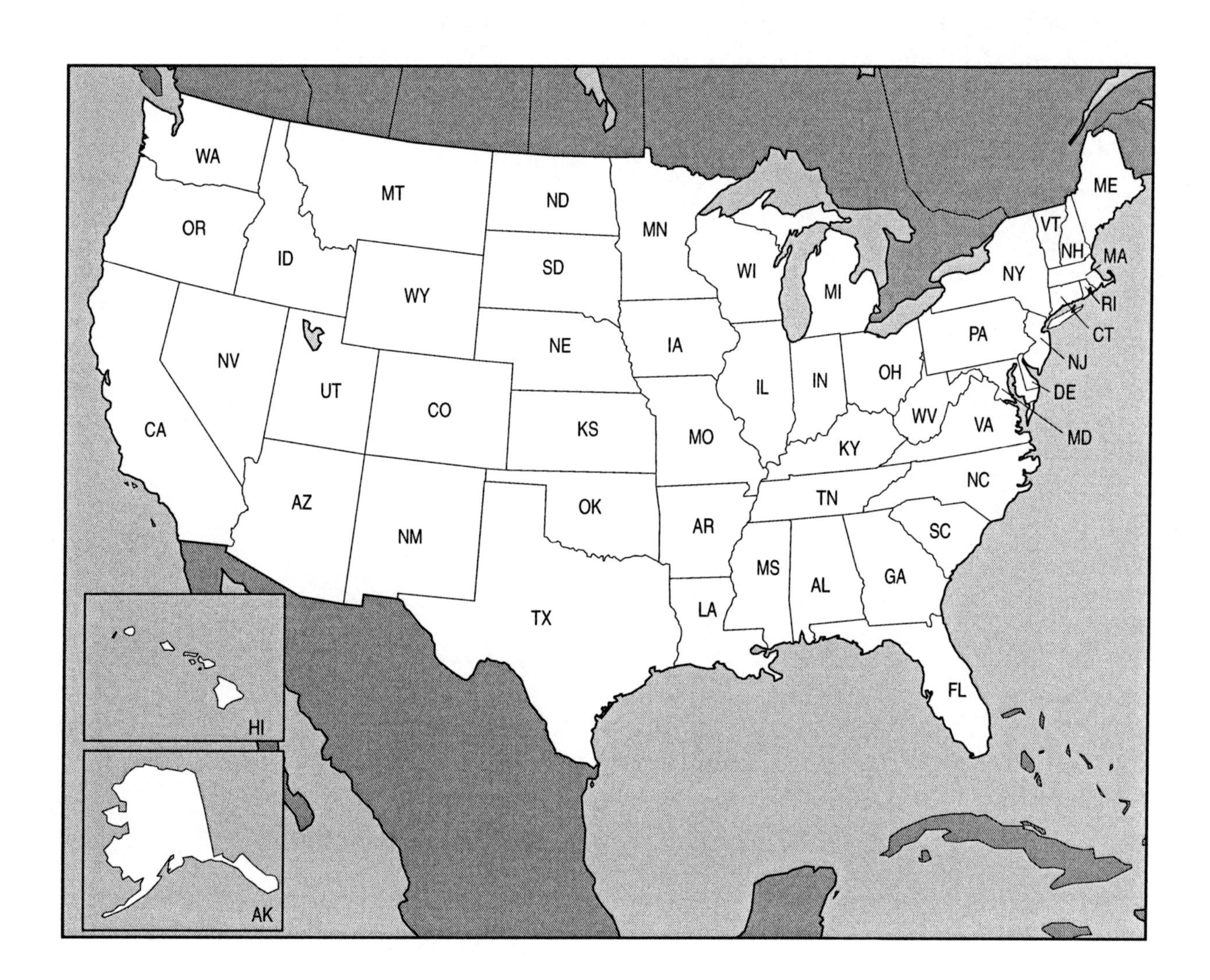

Name________________
Date________________

Australia's Language Policy

In 1987, Australia, a nation in many ways like our own, adopted a National Policy on Languages. The background below on the country and its language policy will help you in a classroom discussion on whether the United States should adopt English as its sole language or pursue bilingualism.

English is the native language of 83 percent of the population of Australia, but the country has higher levels of immigration than the United States. Melbourne is the third largest Greek-speaking city in the world. Large numbers of immigrants from Italy, Germany, Netherlands, China, Poland Malta, Spain, France, and Vietnam have settled in Australia. In addition, Australian Aborigines, or natives, speak about fifty different languages.

The new National Policy on Languages includes four basic ideas:

- Competence in English for *all* Australians
- A language other than English for all Australians, either through the maintenance of existing skills or chances for English-speaking Australians to learn a second language
- Preservation of languages spoken by the Aborigines
- Fair and widespread services in languages other than English, including interpreters, libraries, and media.

Lesson 33
Changing Communities—What Makes the Difference?

Objectives

- To identify the characteristics that determine whether an area is urban or rural
- To illustrate the concepts of rural and urban
- To understand the concept of change—rural areas can become urban areas

Notes to the Teacher

This lesson helps students understand the concepts of rural and urban and apply the terms correctly. Students have difficulty remembering what makes an area urban or rural. Illustrations and simulations help students understand and apply the terms correctly.

The lesson introduction is a book that is normally used with younger children, *The Little House*. Before beginning the story, explain to the class that this book will be used to illustrate some important information that young students would not fully understand. They must listen carefully because the details of the story are important. *The Little House* was written in 1942, but can still be found on the shelves of most public libraries. Some students may even remember the story.

Students brainstorm features that mark whether an area is urban or rural. They create a collage using pictures from magazines to illustrate the terms, rural and urban. Finally, working in small groups, students examine their own community and determine whether it is rural or urban.

Procedure

1. Introduce the book, *The Little House*, to the class. (Virginia Burton, *The Little House*. Boston: Houghton Mifflin Co., 1942.) Explain that this book is usually read to younger students, but there are important ideas in it that those students would find too difficult. Tell students that their job is to listen to the story and to identify all the changes that occur to the little house in the story.
2. Read the story. Have students brainstorm a list of all the changes that happened to the little house. Place the list on the board.

   Suggested Responses:

   House was new and beautiful

   Located out in the countryside

   Lights of the city off in the distance

   Sunshine and starshine visible

   Changing of the seasons

   Country (dirt) roads

   Quiet

   Fields of daisies

   Apple trees

   Built a paved road

   New buildings and gas stations, roadside stands, and small houses

   More roads

   Countryside carved into lots for more houses, apartment houses, tenement houses, schools, stores—Noisy!

   House became old, unwanted

   Trolley cars came

   Elevated train came

   Subway came

   No sunlight or starlight reached the little house

   House became more rundown

   Finally, a granddaughter of the original owners moved the house to the countryside in the middle of a field of daisies with apple trees growing all around. The house could see the sun and the stars; she could watch the changing seasons once again, and she was taken care of.

3. Optional introduction to the lesson if the book, *The Little House*, is not readily available: Have students close their eyes and imagine that they are in a place where they hear the following things: a police siren, lots of cars moving along the highway, the roar of airplanes coming in for a landing, the cries of a hot dog vendor. Ask students where they think they would hear these kind of sounds? (*in the city*) Then ask them to close their eyes again and imagine that they can hear these sounds: the sound of a tractor moving along, birds singing, the sound of hay being rolled into bales, cows mooing. Ask students where they think they would hear these kinds of sounds? (*in the country*)

4. Introduce the terms *rural* and *urban*. Have students define these terms and relate them to the story. How did the story demonstrate rural? (*related to the country*) urban? (*related to the city*) Where was the little house located? (*in the country*) What changed? (*time passed/area no longer rural*) Why do people change rural areas to urban areas? (*Cities grow; population increases; edges of cities move outward to include what had once been rural*) What causes such a change? (*Population shifts into rural areas and grows large because of economics—industries are started in an area, and people come to work in those industries and the population grows; a rural area may lie on a crossroads and people move there to take advantage of this movement and trade opportunities; an area may become a trade center.*) How long did this change process take? (*two generations*)

5. Supply each student with two sheets of posterboard, 9" by 12" or 12" by 18", and a supply of old magazines. Working in groups of four, have students label, with colored marker, one sheet of posterboard, Urban and the other Rural. Cut pictures from magazines to illustrate the terms. Display their completed illustrations and discuss their insights.

6. Use the same groups of students to begin a research project about the local community. Ask students to find out all they can about the community a hundred years ago (longer, if desired) and compare it to the community of today. Was it ever rural? Is it still rural? Is it urban today? If so, what caused this to change? If possible, find pictures to compare and discuss how and why communities change over time.

7. Have groups present their findings using visual aids and any other creative means they can devise to make the presentations exciting and interesting.

8. Be sure students understand that the shift from rural to urban is common to industrialized countries around the world and not just an American phenomenon.

Enrichment/Extension

1. Research your state capital city. Compare it as it is today with the way it was when it was established. How has it changed? Did it ever have characteristics that were rural?

2. Design a community based on the following set of information. Give reasons for your design. Structures should have a specific function. Be prepared to share your planned community and explain it to classmates.

   This rural area has only three wheat farms located in the region—spaced some distance apart. The large city that is forty miles away has decided to build an international airport in that general area. People and businesses will be starting to move out there from the city. Design the new community that will grow up as a result of this expansion. Think about all the things that will be needed in an urban area—water service, fire and police protection, electricity and gas, grocery stores, banks, etc.

Lesson 34
Sports—An Aspect of Culture

Objective

- To consider sports as a cultural trait

Notes to the Teacher

This lesson has students look objectively at sports as a cultural trait. Just as literature, clothing, and folkways speak about the character of a culture, so too can sports. Sports also show the diffusion of culture, that is, the spread from one culture to another of a particular sporting event. In addition, sports demonstrate the interdependence of nations, as they compete internationally, as in the modern Olympics or the Pan-American games. The worldwide television viewing audience of these games attests to the ability of sports to bridge both geographic and ideological differences between nations of the world.

The probable motive behind the origin of various sports varies widely. The following is a partial list:

1. Young men in training
2. Children emulating adults in hunting or warfare
3. Warriors or hunters practicing
4. Intertribal or interclan competition for prizes, land, or women
5. Single combat of heroes being mirrored
6. Praise or social acceptance
7. Mythological representatives
8. Ceremonial functions for victory, funerals, holidays

Historically, the earliest records of sports come from Egypt, from paintings on tomb walls (2300–2100 B.C.) showing Pharaohs, priests, and women engaged in wrestling, boxing, fencing, swimming, rowing, running, jumping, weight lifting, archery, and horseback riding. In ancient Ireland, accounts of the Tailteann Games in 1800 B.C. record events in running, jumping, wrestling, horse racing, and hammer-throwing. The most famous historical sporting events are the ancient Greek Olympics, documented as being held every four years for ten centuries, from 776 B.C. to 393 A.D., involving track and field events, wrestling, boxing, horse racing, and field hockey. The Greek Heraean Games were organized for women to compete in footraces, wrestling, and chariot racing.

Students prepare a map showing the prevalence of several major sports around the globe. They use their completed maps as a resource for a class discussion of sports as a cultural trait. To conclude the lesson, students write a paragraph on the thematic question: How do sports interest and participation in the United States reflect this country's cultural traits?

Procedure

1. Begin by asking the following questions:
   a. How many of you have gone to an art museum or a performance of folk dances in the past year?
   b. How many of you have gone to or participated in a sporting event in the past year?
   c. If art and dance are cultural traits, could sports also be considered a cultural trait? Why, or why not?

   Help students to understand that any aspect which reflects cultural interest and participation is a cultural trait.
2. Share the information in the Notes to the Teacher on the origins of sports to provide background.
3. Distribute **Handout 83** and have students complete the map activity as indicated. Consider having students work in pairs to complete the maps. (A completed map is provided at the end of the teacher section.)

4. Have students refer to their completed map as they consider the following questions in a large-group discussion of sports as a reflection of a nation's culture.

   a. What differences in sports participation do you note between richer and poorer nations? How do you explain those differences? (*Richer nations participate in more organized sports than poor nations. The cost of training and equipment, greater leisure time, and proximity to urban sports arenas account for differences.*)

   b. What differences do you notice between countries that have large cities and those that are largely agricultural? Do you suspect that the rural areas have no interest in sports or that they engage in different sports from the ones on your map? Explain. (*Countries with large urban centers have more spectator sports; running, however, is universal because it can be done anywhere and requires no expensive equipment or training, at least on the amateur level.*)

   c. What Olympic sports have you seen on television that would be ideal for individuals with little money for equipment? What individual sports require costly equipment? (*Running, jumping, and swimming require little equipment. Golf and polo, however, are often viewed as sports of the rich because of the equipment and training needed.*)

   d. In what ways does climate affect the sports interests of a nation? (*Winter sports are popular in colder climates, for example bobsledding, skiing, ice hockey, or the luge are not possible in warmer climates. On the other hand, golf and baseball are less common in areas with short summers.*)

   e. What factors might help to explain why some nations put a greater emphasis on individual rather than team sports? (*Team sports require arenas or stadiums, proximity to urban centers, and money for training and equipment.*)

   f. In what ways do countries link patriotism and sports? (*Many nations sponsor training for their athletes in order to project a favorable image and bring national recognition at the Olympics and other international competitions.*)

   g. How do sports reflect a general aspiration to excel? (*Sports offer opportunities for individuals, teams, and countries to push to be the best that they can be and to demonstrate their superiority over others in acceptable and nonviolent ways.*)

5. In what ways does your completed map suggest the influence of each of the following factors in a country's sports interests? Give at least one illustration for each.

   a. Leisure time (*Less affluent countries of Africa and Asia participate less in organized sports.*)

   b. Technology (*Countries with widespread access to television, for example, commonly have a wide range of sports interests.*)

   c. Climate (*People living in warm climates cannot easily participate in outdoor winter sports such as bobsledding or skiing.*)

   d. Availability of arenas and other sports facilities (*Sports facilities promote team sports.*)

   e. Migration of people from one place to another (*The United States, a "melting pot of nations," has "adopted" many sports that originated elsewhere. Also, sports popular in Great Britain are also popular in the Commonwealth countries.*)

   f. Widespread availability of public transportation (*Access to public transportation increases access to large arenas and spectator sports.*)

   g. Media coverage (*Media coverage has introduced new sports and increased people's knowledge of sports possibilities. Much of the increased interest in the United States for gymnastics results from widespread media coverage.*)

6. Conclude by having students work in pairs to write a paragraph on the following question: How do sports interest and participation in the United States reflect this country's cultural traits? Share and discuss the paragraphs. (*Possible factors reflected in student paragraphs are traditions, leisure time, proximity to urban centers, immigration patterns, media coverage, climate, and public support for sports in schools and colleges.*) Provide time to share the paragraphs and discuss.

Enrichment/Extension

1. Research the origins of various sports by answering the following questions:
   a. How did the sport originate?
   b. When and where did it begin?
   c. Where and when is it played today?
   d. How do you account for its popularity?
   e. In a paragraph, explain the main point of the contest.
   f. How does it reflect the culture of countries where it is played?
2. Research the sports interests of other countries. Write a report on your findings which includes information on at least five other countries.

Suggested Responses, Handout 83:

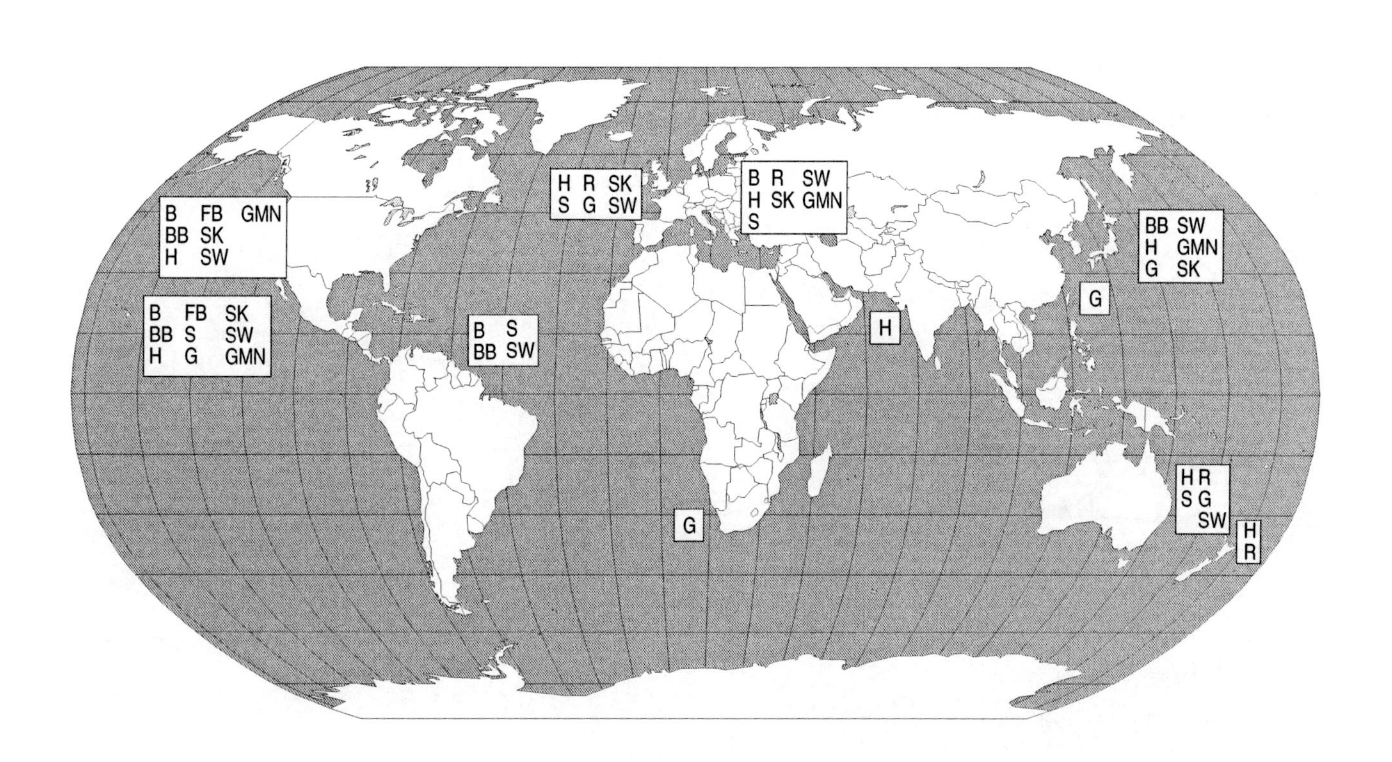

Name________________
Date________________

Sports—An Aspect of Culture

Complete the accompanying visual to use as a resource for a class discussion on sports as an aspect of culture. On the map, write one or more letters in the large boxes to indicate sports which are popular in each area of the world. Use the following key:

BB Baseball	**FB** Football	**R** Rugby	**SK** Skating	**SW** Swimming
B Basketball	**S** Soccer	**G** Golf	**H** Hockey	**GMN** Gymnastics

The following lists of areas where various sports are common will provide necessary data.

Baseball—Latin America, Puerto Rico, Japan, United States, Canada

Basketball—United States, Canada, Mexico, Poland, France, Brazil, Russia, Uruguay, Argentina, Italy, Puerto Rico

Football—United States, Canada

Soccer—Great Britain, Australia, Europe, Latin America, United States

Rugby—Great Britain, France, Australia, New Zealand

Golf—United States, Great Britain, South Africa, Australia, Japan

(Ice) Skating—United States, Great Britain, Europe, Japan, Canada

(Field) Hockey—United States, Canada, India, Netherlands, Great Britain, Germany, Pakistan, Australia, New Zealand, Japan, Spain

Swimming—Great Britain, Europe, Japan, China, Canada, United States, Australia, Mexico

Gymnastics—Germany, Switzerland, Eastern Europe, China, Japan, United States, Canada

Be sure to complete the key to the map.

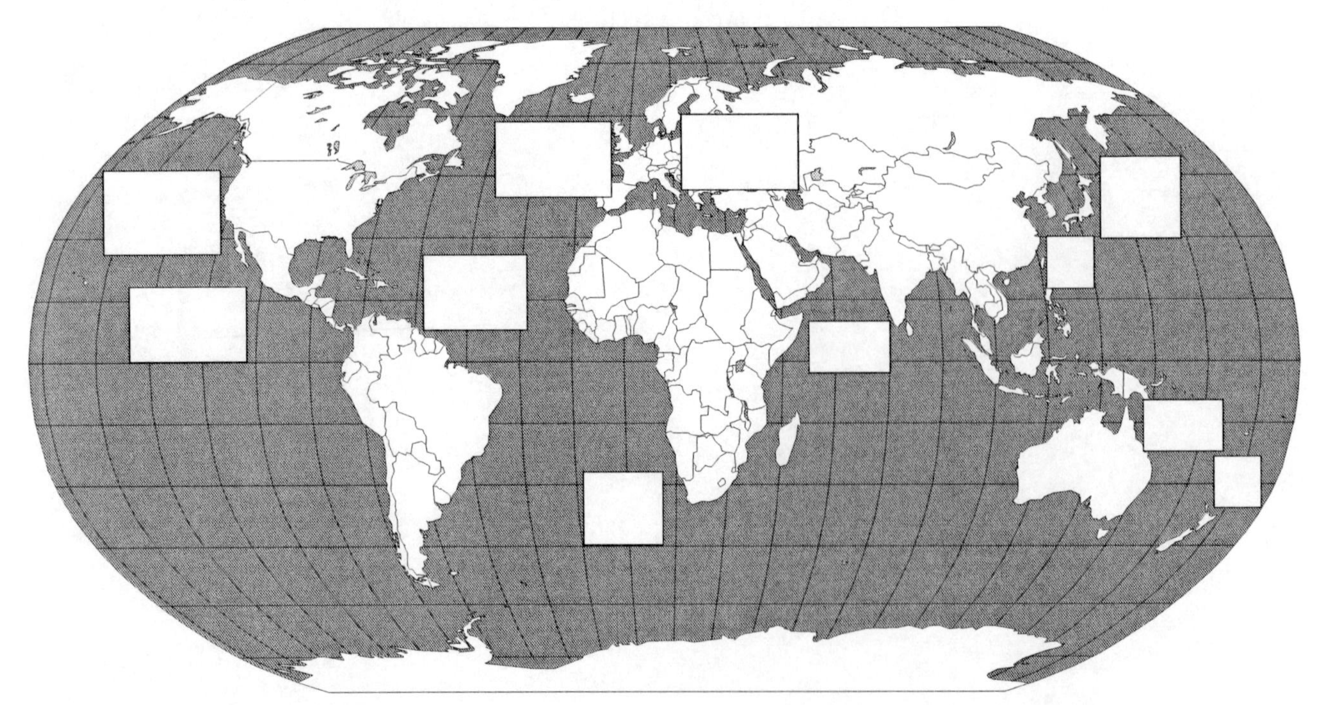

Lesson 35
Technology—Its Cultural Impact

Objectives

- To describe changes brought by recent advances in technology
- To explain the transmission of technology from one region to another

Notes to the Teacher

The kind of soil, climate, natural vegetation and animals, as well as the presence of minerals, influence the nature and extent of people's achievements within each region. The extent to which people utilize their natural resources is related to their desires and to their level of technological understanding.

Technology has put new and powerful instruments into the hands of people all over our global community. These new technological advancements have allowed people to explore their environments and discover secrets of outerspace as well as what lies beneath the seas and the surface of the earth. The complexity of technology can be observed through powers being generated by falling rain or by the releasing of nuclear energy, the cooling and heating of living spaces, or communication and transportation systems which allow people to have contact with people in other parts of the world.

Students poll their parents and/or adults to find ten inventions, discoveries, or medical advances made during the adults' generation that have made a significant impact on their lives or shaped their thinking. Students then research individuals or countries responsible for these innovations. In class, students make a composite list of the advances and brainstorm their consequences—both good and bad. After discussion of the impact of technology and its spread throughout the world, students conclude by citing specific examples of several generalizations about the impact of technological development in our society.

Procedure

1. Several days before discussion of this lesson, distribute **Handout 84** and assign the poll and related research as a homework assignment.
2. On the day of discussion of this lesson, explain technology and its relationship to science. Point out that technology is the use of science and machines to solve practical problems.
3. Have students make a composite list of the inventions, discoveries, and medical advances adults mentioned in the student polls. Ask students to identify the practical problem solved in each case. Let them share personal stories of adults they interviewed on the impact of changes on their lives and thinking. Have students place colored pins on a large world map to show the country primarily responsible for each innovation.
4. Have students make generalizations about the sources of change today. (*Movies, television and computer sources—Internet, The Web—are possible student responses.*)
5. Explain to students that improvements in technology often depend upon cultural contacts as cultures borrow, change, and use technologies from each other. Point out that technology has made it possible for countries to import and export products from each other. Ask students why countries would want to import and export from each other. (*Countries import goods that they do not produce and they export goods to other countries that need their goods. The exchanging of goods can improve the economy of the countries involved in exporting and importing.*)

   Discuss ways in which the United States both imports and exports technology. (*Most VCRs are imported from Japan; American airplanes constitute the world's airlines.*)

6. After the students have completed their composite list, explain issues such as the following:

   a. The development of the steam engine was the major event that revolutionized transportation in the nineteenth century. What evidence can you cite to support that statement? (*Growth of railroads and steamships, etc.*)

   b. What invention in the twentieth century has done the most to revolutionize communication? What evidence can you cite to support your contention?

   c. In what ways has the world changed since you were born? (*Students should be able to cite technological changes, such as space travel, space shuttles, communications satellites, and computers, as well as changes in the world order*—e.g., *the breakup of the Communist empire—that have vastly altered the world in which we live.*)

7. Conclude the lesson by having students work in small groups to complete **Handout 85**. Allow class time for students to share their responses. This discussion should summarize their learning.

Enrichment/Extension

1. Research information on the people listed. Write a report which tells how their inventions influenced technology.

   a. John Kay

   b. Robert Fulton

   c. Eli Whitney

   d. Michael Faraday

   e. Cyrus McCormick

   f. Edwin H. Land

2. Use magazine and newspaper articles to start a bulletin board on technological changes occurring in the world today. Highlight your articles with pictures.

3. Make a list of things that have improved your school and home environment. Write a paragraph or two explaining how your life would be different if we didn't have these modern conveniences. Share paragraphs orally or place on the bulletin board so others can read them.

4. Invite a speaker from a computer center to discuss the impact computers have had on technology and American life.

Name________________
Date________________

Technology Affects Our Lives

Recent inventions, discoveries, and medical advances have had a significant impact on people's lives and ways of thinking about the world in which we live. Many of these innovations did not exist when your grandparents or even your parents were in school. For homework, poll several adults to compile a list of ten inventions, discoveries, or medical breakthroughs that have had a significant impact on their lives or thinking. Ask each person to state how each change affected him or her personally. Then try to research the individual(s) or country responsible for each of these changes. Your completed poll and research will be an important resource for a class activity on the effects of technology.

Poll Results

Invention, discovery, or advance	Impact	Person(s) or country
1.		
2.		
3.		
4.		
5.		
6.		
7.		
8.		
9.		
10.		

Name________________
Date________________

The Impact of Technology

Work in your small groups to cite three pieces of evidence to support each of the generalizations below about the influence of technology today.

1. Technological advances today do not come from any one person or country.

2. Modern technology generally improves life compared to the past.

3. Technology has created new problems as well as solving old ones.

4. Robot technology is increasing in importance.

5. The United States both imports and exports products important to technology.

6. Technology has had a great impact on the lives of individuals.

7. A new invention often causes mass production of others.

8. Technology has brought changes in people's thinking about the world in which we live.

Lesson 36
Education—
The Transmission of Culture

Objectives

- To understand education as an enculturation process
- To compare values reflected in the schools of three countries

Notes to the Teacher

Human society has always found ways of transmitting the values, mores, norms, and traits of the culture to its young. In primitive societies, the process of transmission was accomplished through individualized training or, at most, through small-group instruction. As societies have progressed and become more complex, the methods of cultural transmission have become increasingly specialized. Education is a major means of transmitting culture, of making young people acquire the traits, values, and norms of their culture—in short, education is a process of enculturation.

In primitive societies, young men were often trained in the skills of hunters or warriors by a male family member, while young women were trained in domestic arts by the older women in the family. In more advanced societies, as in medieval Europe, class-structured societies had specialized training and/or schooling for nobles, religious, military, and merchants and little, if any, education for peasants. Craftsmen and merchants studied through apprenticeships; people learned, for the most part, stories of their culture through storytellers or the family unit; scholarship was limited to clerics or nobility. Modern societies indicate one cultural value by emphasizing compulsory, equal opportunity education for the young.

Today young people learn cultural values through many avenues. Family values—and often ethnic and religious values as well—are learned at home. Religious values are often acquired through the religious institution a person attends. Educational and cultural values are most often taught by the schools. Peer groups and media also express values regarding certain aspects of one's culture.

This lesson focuses students' attention on what is meant by the concept of education as a process of enculturation. To begin, students brainstorm a list of positive and negative characteristics of American education and record values these features reflect about our society. They then read a short description of Japanese education and take a look at the seventh-grade curriculum in Greece to see values reflected by the educational systems in those two cultures. To conclude the lesson, students consider the impact on society—both for good and ill—of educational systems in the two countries and prepare a list of "lessons" America might learn from the examples of education in Japan and Greece.

Procedure

1. Ask students to generate a list of positive and negative features of our American educational system. (It may be useful to have in class several newspaper or magazine articles on the state of American education today.)
2. Ask students what American values their lists suggest. (They may list, for example, low priorities on funding education, equality of sexes, inequality of opportunity for students in some locations or classes, too little emphasis on excellence, little priority on foreign languages, expectation that schools will solve social problems, and acceptance of failure.)
3. Assign students to read **Handout 86** in class or for homework.
4. Ask what values are reflected in the practices of Japanese and Greek schools.

Suggested Responses:

Japanese schools—

1. *Education as a national priority*
2. *Expectation that all can learn*
3. *Discrimination between sexes acceptable*

4. *Large degree of conformity*
5. *Pressure to succeed*
6. *Emphasis on rote learning*

Greek schools—

1. *Modern and ancient Greek heritage and language*
2. *Ethnocentric cultural and historical pride*
3. *Religion as high value in Greek culture*
4. *Equality of educational programs for sexes not valued*
5. *Geology, a subject useful in a farming economy*

Ask what impact these values have on their individual societies, both for good and ill? (*Students might mention, in the case of Japan, factors such as strong pressure for quality, lack of emphasis on creativity, loss of much talent of women in business, belief that individual effort pays off, and the idea that failure at the bottom is neither acceptable nor necessary.*) In the case of Greece, schools stress the value of religion; skills and understanding necessary for farming, which is key to the Greek economy; failure to require the most from girls; strong tendency to promote understanding of the cultural heritage of the nation's past.)

5. To conclude the lesson, ask each student to write a list of top five "lessons" of either a positive or negative nature that the United States might learn from the experiences of the Japanese and Greeks in inculcating the values of the society through education. Share and discuss the listings.

Enrichment/Extension

1. Find articles on education in other countries, for example Russia or Germany, to learn what values of their societies are reflected in their educational systems.
2. Interview students in your school who have attended schools in other countries to learn differences in educational practices and what this says about cultural values in those countries. Invite these students to address the class.
3. Write an essay explaining what one practice of Japanese or Greek schools would most improve the American educational system and give your rationale.

Name________________
Date________________

Japanese Education

Read the following description of Japanese education and study the seventh-grade curriculum for Greek schools in preparation for a discussion of education as enculturation in those countries.

Japanese schools convey a clear message to students that they are there to learn, not to socialize, and that with effort and persistence all students can master all the required material. The school week is five and a half days, with the half day coming on Saturday morning. The school year in Japan is 240 days, rather than the 180 days typical of American schools. Japanese students have two vacations of three weeks each, one in the spring and one in the summer. In addition, by age eight or ten nearly all Japanese students have private tutoring or attend evening "cram schools" to prepare for the demanding and all-important entrance examinations for high school and college. Scores on these tests are virtually the only factor considered in determining which students will attend the best schools.

Discipline in Japanese schools is strict and begins early. In "Nude Kindergartens," young boys go shirtless and wear short pants all winter in order to build endurance. Older students wear uniforms. Most schools do not permit girls to dye or perm their hair or wear perfume or makeup. Boys are not permitted to wear their hair below their ears or collars. Rigorous coursework does not allow time for after-school jobs, and Japanese students are not allowed to go to coffee-houses, discos, etc. The motto for many regarding hours of sleep is, "Pass with four, fail with five."

Coursework, too, indicates Japanese priorities. Until 1992, there were no electives, and all students were expected to complete twelve years of science and math, including chemistry, physics, and calculus. Fewer than half of American students take either chemistry, physics, or math, and only 6 percent take calculus. As a result, Japanese students complete high school three or four years ahead of most American students. Very recently Japanese schools introduced new courses to focus on independent learning and problem solving rather than a sole dependence on teacher lectures and student notetaking.

As in much of Asia, expectations for boys and girls are different. Japanese women are expected to be committed to their children's welfare, and girls often sacrifice their own career aspirations in order to allow for a better education—and career opportunity—for their brothers.

Name________________
Date________________

Greece—7th Grade Curriculum

Subject	Hours per Week
Religion	2
Ancient Greek	4 (9 hours total Greek)
Modern Greek	5
History	3
Mathematics	4
Foreign Language	3
Geography/Geology	1½
Biology: Botany and Zoology	1½
Music	1
Arts	2
Physical Education	2
Technology (boys)	2
Home Economics (girls)	2
	33 total hours per week

Lesson 37
Leadership—
Women Who Have Made a Difference

Objectives

- To identify and examine the qualities of selected successful women
- To expand research and presentation skills

Notes to the Teacher

Students examine the characteristics that allow people to gain success—with women serving as the focal point. Many women have been successful in various fields—politics, science, law, writing, education, etc.—but they are often not highlighted for students. It is important for young women to see successful role models in all areas of pursuit. Women have been leaders in various fields both in the past and at the present time. Males need to understand that females can be successful in any area that they wish to study. It is vital that students do not see their roles as gender specific.

Students read about several women and their successes. Discussions about the problems that they had to overcome to become successful follow. Students are given a selection of women to research, and a related project is completed. Students make a presentation to the class upon completion of the project.

Procedure

1. Pass a Susan B. Anthony dollar coin around the class. Survey to see how many recognize her. If anyone knows, let them tell about her; otherwise, tell the class about her. Use the following information:

   Susan B. Anthony

   She was an abolitionist, a feminist, a suffragette, and a rebel. She broke the laws of New York, when she went to the polls and voted in an election. She was a woman and not allowed to vote according to state law. She was arrested, tried, and found guilty. She told the court that she would never pay the fine—and she never did. (November 5, 1872)

   She was a teacher. She spoke out against injustices against women. She wanted equal pay for women teachers, coeducation, and college education for women. She joined Elizabeth Cady Stanton and Lucretia Mot, two brave feminists, in their fight for women's suffrage. The public and the press did not like the idea of women voting. Susan became the leader of New York's Women's Rights Association. She tried to get several reforms passed: the right for women to vote, the right for married women to handle their own money, and the right to equal guardianship for their children. Finally, the legislature (all men) did pass a bill (1856) that allowed women to control their money and to have equal guardianship of their children.

   She spoke out against slavery as the Civil War approached. After the war she lobbied for the passage of the Thirteenth Amendment which abolished slavery. She was bitter about the passage of the Fourteenth Amendment, not because it allowed the ex-slaves to vote, but because it said that only *males* could vote. She continued fighting and encouraging more and more women to join her movement for women's rights. She retired at age 80. She died in 1906. Her dream was realized after her death. In 1920 the Nineteenth Amendment was added to the U.S. Constitution and gave women the right to vote. Without Susan B. Anthony and others like her, that day may have taken even longer to arrive.

2. Ask students to brainstorm some of the traits they think Susan B. Anthony had that made her successful. (*perseverance, bullheaded, independent, intelligent, fighter, etc.*)

Distribute **Handout 87** and allow time for students to read about these women. Discuss what traits each of them had that allowed them to be successful. Chart them on the board and mark those that are the same. Ask students if they share any of these traits. Follow with questions like these:

a. What common thread of experience runs through each story? (*They are all hard-working, goal-oriented women.*)

b. What types of problems did each face?(*G. Meir—belonged to an ethnic group that many discriminated against; living conditions were not easy; I. Gandhi—family fought against the ruling government and were often involved in political struggles; M. Thatcher—struggled to gain a political office after being defeated and popular opinion was frequently against her; S.D. O'Connor—struggled to balance her family and her work, had to finally take time off and stay home to care for three children.*)

c. What obstacles did each face? (*G. Meir—had to work at many different types of jobs and give up being a teacher in America to live in a place where Jews were not hated; I. Gandhi—had to overcome her own personal illness, the death of her husband when she was only 43, and the political upheaval in India during the 1970s; M. Thatcher—served as prime minister at a time when Britain's economy was at its lowest; she wasn't always successful in trying to improve the economy; the public didn't always like the way she chose to spend money on the military; S. D. O'Connor—greatest obstacle to getting on the Supreme Court was the fact that she was a woman*)

3. Distribute the list of women's names on **Handout 88**. Have students choose one name on the list to research and do a project about. Allow no repetition of names until every name has been chosen. Have students complete the activity as directed. Encourage students to draw or copy a picture of the subject. Consider doing this activity in a pair format.

4. Review the project directions carefully.

   Project Directions: Get a project board or create one (a three-sided, free-standing upright board with the wing sides of equal size.) Using the information gained through research, copy or draw pictures of the person being researched. Mount the pictures and printed information on the board (not the whole research record—just significant highlights.) There should be a title across the top in large cut-out letters.

5. Display presentation boards. Each student (or pair of students) should present a short talk about the individual chosen and use the board as visual aid.

Enrichment/Extension

1. Present your research data by role-playing a television talk show where you take on the roles of the individuals you researched. Be sure to designate a student to play the host. Try to keep each role play to a maximum of six "guests."

2. Create a multi-tier timeline showing when these women lived and what major events were going on in the United States and the world during their lifetimes. Display your completed timelines.

Name________________
Date________________

Women Who Made a Place for Themselves

Golda Meir

Who could imagine that little Goldie Mabovitz (Mabovich) would grow up to become the prime minister of Israel? Certainly not her as she made mud pies with her friend near her home in Pinsk in Kyyiv, Russia. She only knew that the Russians were cruel to the many Jews who lived in her town. The Russian Cossacks rode their horses recklessly close to Golda and her friend. As they and their huge horses flew over the two little girls, Goldie must have thought that she too would become one of their many victims.

Goldie, known today as Golda, was born in Kyyiv on May 3, 1898. (This part of Russia has taken back its old name and today we know it as the Ukraine.) In 1905, her father immigrated to Milwaukee, Wisconsin, and his family followed him the next year. Throughout her life, Golda set high goals for herself and others. When she was a fourth grader she collected money to buy books for poor children. She became a Zionist (one who believed that a Jewish nation should be established in Palestine) when she was a teenager.

Golda attended the Milwaukee Teachers College and married Morris Meyerson in 1917. Four years later, she and her husband immigrated to Israel to live on a kibbutz (a collective farm). At this time, Golda Meyerson changed her name to Meir, the Hebrew form of Meyerson. She was determined to work at any job she was offered in order to make her dream of a Jewish homeland come true. She wanted all Jews to live in a place where they would not be hated.

Golda Meir was active in the Zionist groups in Palestine, Europe, and the United States in the 1930s and 1940s. When Israel declared its independence in 1948, she was one of the signers of the proclamation. Later she served as a foreign minister and minister of labor. Finally, in 1969, she was elected prime minister. She resigned in 1974, because of differences of opinion about Israel's readiness for the Yom Kippur War of 1973.

She died in Jerusalem on December 8, 1978, knowing that her beloved country, Israel, was well established. Her countrymen and the world mourned her death.

Indira Gandhi

It is hard to imagine what life must have been like for little Indira Gandhi as she grew up in India in the 1920s and 1930s. Her father, after all, was Jawaharlal Nehru, who led his people against the British who then ruled their country. This was a revolution, and even Indira had a part to play in it. She lead a group of children called the "Monkey Brigade." They ran errands, helped spin cotton into thread, and made red and green Indian flags which they displayed everywhere they could. The police could never seem to catch them at these deeds even though they were close by. Why the children even jumped rope and played games near the police station so they could pick up news about police plans to arrest the rebels. Then, they hurried back to the victims with the information so they could escape.

Indira Gandhi was born in Allahabad on November 19, 1917. She was an only child. She graduated from Visva-Bharati University, Bengal, and she also attended Oxford University in England. In 1938, she became a member of the National Congress Party and began to work hard to gain India's independence from Britain.

She became quite ill and had to go to a sanitarium in Switzerland for a few months. When she came home, she told her family that she was going to marry a young Parsee (an Iranian), but they strongly disapproved of the

Name________________
Date________________

match. Because her father had always believed in equality regardless of religion or caste, he finally had to give her his blessing. In 1942, she married Feroze Gandhi, a lawyer, who was an active party member like herself. Not long after their wedding, they were both arrested by the British for their secret activities and kept in prison for thirteen months. They had two sons, but her husband had two heart attacks and died leaving Indira a widow at the age of forty-three.

She helped her father until his death in May, 1964. She remained politically active and was finally elected prime minister of India in January, 1966. She lost her seat in parliament in 1977, but she made a surprising comeback in 1980 when she formed a new majority government. She tried to stop the Sikhs, a religious minority group, from rebelling. She was criticized all over the world because of the way she suppressed these people. On October 31, 1984, she was shot and killed by two of her bodyguards who were members of the Sikh group. Her son Rajiv became the new prime minister and served until 1989. He was killed on May 21, 1991, by a bomb during an election rally in Madras.

Margaret Thatcher

Everyone likes to win, and Margaret Thatcher was no exception. There was always a challenge to be undertaken, and again and again she proved that she was equal to that challenge. In high school she became the youngest captain the school ever had on the field hockey team, and later when it came to negotiating a hard line, even the Russians respected her. They gave her the nickname "The Iron Lady."

Margaret grew up in a hard-working family. Her father ran a grocery store in Grantham, and they lived in an apartment over the store. Her father was interested in politics, and Margaret grew up listening to his discussions. She, too, came to be interested in politics.

She was born in 1925, in Grantham and went to college at the University of Oxford. She earned degrees in chemistry. She worked as a research chemist for four years after graduating. In 1953, after studying for the bar, she became a tax lawyer.

Thatcher joined the Conservative Party and won a seat in the House of Commons in 1959. Again, she was the youngest woman ever to try for a seat in the House of Commons. She had married Denis Thatcher, a businessman, and they had a set of twins, Carol and Mark.

In 1979, she became the prime minister of England, the first woman to ever hold this office. She remained in office until 1990. She took office when England's economy was in trouble, their unemployment was at its highest, and the spirit of the people was at its lowest. Some of her efforts to make changes were successful while others were not. When she sent troops into the Falkland Islands, many people criticized her. She was blamed for putting too much money into a modern navy with nuclear subs instead of the traditional ships, but when the troops were victorious in the Falklands, people's opinion of her naval policy changed. Margaret Thatcher is a doer—she gets the job done.

Today Margaret Thatcher is a private citizen. Since resigning from the office of prime minister, she spends time with her family, working with various charitable organizations, and visiting places to give speeches about politics and her career in public office.

Name________________
Date________________

Sandra Day O'Connor

The news was exciting! The first woman to serve on the Supreme Court was selected in 1981—Sandra Day O'Connor. Who could imagine that a little girl born in El Paso, Texas, could ever get to the U.S. Supreme Court?

She rode her first horse at age three, and as she grew older she learned to brand cattle, mend fences, ride horses, and drive a tractor. She did all these things during her summer vacations. She graduated from high school at sixteen and went to Stanford University in California to earn a law degree. No one would hire her in the early 1950s because there were few women lawyers and people thought that women could not be good lawyers.

She married John Jay O'Connor in 1952. She and her husband worked as lawyers in West Germany. In 1957, they moved to Phoenix, Arizona, where they had their first son. Later there were two more children. For a while she stayed at home full-time to take care of her children. She went back to work in 1965. She became an assistant attorney general in Arizona and later a state senator. In 1974, she became Arizona Superior Court Judge. People always noticed how fair she was—she enforced all laws, even those she did not like.

In 1981, President Ronald Reagan appointed her as an associate justice to the U.S. Supreme Court. Congress approved her appointment and she will continue to serve as a Supreme Court Justice until her resignation or death.[1]

[1]Adapted from B. T. Keller, C. Shaerfi-Murphy, J. L. Pavia, and K. M. Tryda, *U.S. Biographies, Book 7 1960–1990* (Villa Maria, Pa.: The Center for Learning, 1992,) 99–101.

Name________________
Date________________

Women Who Have Made Contributions

Select one name from the list. Research and write a report that includes background information about the person you have chosen, what goals she set for herself, what obstacles she met, and how she overcame the obstacles to become successful.

Create a project from your research. Draw or copy pictures of the person and events/details in her life. Choose a title for your project and mount pictures and bits of information collected. These should be mounted on a three-sided, free-standing presentation board. The board should be neat, pleasing to the eye, and tell at a glance the story of this successful person's life.

Be ready to present your board and a brief oral summary of your research to the class.

Be sure to cover these points:

1. Who is/was she? What qualities of leadership does/did she show?
2. What is/was her field of work?
3. What are/were her accomplishments?
4. What obstacles did she overcome? Did/do people criticize or admire her?
5. What conditions in her society led her to do her kind of work?

Choose One

Abigail Adams
Susan B. Anthony
Corazon Aquino
Elizabeth Blackwell
Margaret Bourke-White
Rachel Carson
Shirley Chisholm
Marie Curie
Mary Harris Jones (Mother Jones)
Margaret Mead
Maria Montessori
Toni Morrison
Sandra Day O'Connor
Rosa Parks
Eva Peron
Leontyne Price
Sally Ride
Eleanor Roosevelt
Mother Teresa
Valentina Tereshkova

Lesson 38
Statistical Information—Median Ages of Nations

Objectives

- To examine the concept of median ages of nations
- To complete exercises in statistical classification and interpretation

Notes to the Teacher

The median age of a nation is the approximate age, determined by average, showing that half of the population is over the median and half of the population is under the median. Nations with high median ages have older populations and special future concerns, such as, care for the elderly, changes in the job market (often more service industries), and health and housing. Nations with low median ages have younger populations and special future concerns, such as housing concerns and providing jobs, educations, and training for a growing work force.

Traditionally, agricultural societies valued large families to support the farming industry, leading to nations with low median ages. Modern industrialized nations now show a concern for smaller families and slower growth rates that are more economically manageable. Concepts such as "zero population growth," *i.e.*, two offspring to replace the two parents, have arisen and have gained support in some industrial nations. China has embarked on a government sponsored emphasis on only one child per family among the Han majority (but not the ethnic minorities) to curb the population. Many nations, due either to economic necessity or religious beliefs, continue to experience rapid population growth.

The focus of the lesson is for students to work with statistics as a means of categorizing nations and their cultures through analysis of median ages. Some of the terms may need to be either introduced to students or reviewed in order for them to succeed with the statistical emphasis of this lesson.

Terms

Median age—the average of middle statistic of ages, showing that half the population is above, and half the population is below, the average age

Life expectancy—the average age at which people die, *i.e.*, the age to which a person can expect to live

Per capita income—the average earned wage of a person in a nation

Annual percentage of population growth—the percentage of growth or increase per year in a nation's population

Literacy rate—the percent of the population who can read and write

Procedure

1. Introduce the topic using material in the Notes to the Teacher. Stress the definitions of key terms. Place the terms on the chalkboard and help students reach consensus definitions.
2. Distribute part A of **Handout 89**. Review the chart before part A to make sure that students understand the terms of the characteristics and the concepts that they represent. Make sure that students realize that these characteristics are generalized, and that there are specific instances that may not match these general categories exactly, but that they show exceptions to the categories. Discuss the United States as a high median nation. The United States has 22 percent of its population under 15 years, 13 percent above 65, a life expectancy for males of 76 years, a 7 percent annual population growth rate, and a 97 percent literacy rate.
3. Have students complete the activity in part A as an individual assignment. Review their responses and their reasoning in each case. Have students locate these countries on a classroom map.

Suggested Responses, Part A:

1. *a. low*

   b. low

   c. high

   d. low

   e. low

   f. low

   g. low

   h. high

   i. high

   j. high

2. *a. low*

   b. low

   c. high

   d. high

   e. low

3. *It is reasonable to expect Chad to be poor, underdeveloped, and tropical. Life expectancy, literacy, and per capita income are all likely to be low.*

4. Part B of **Handout 89** is a concluding activity. Divide the class into small groups (three to five students). Distribute part B. Have half of the class fill in the blank with "high" and half of the class fill in the blank with "low." Allow each group ten to fifteen minutes to identify the problems and suggest possible solutions. Have groups select a leader who will report the findings to the class. For consistency, begin with all the high median or all the low median, then have the opposing category groups report. Consider having some or all of the responses listed on the chalkboard or on an overhead projector.

Suggested Responses, Part B:

Responses will vary, however, certain answers may occur.

High median—*maintain job market, retirement, filling jobs, medical concerns for elderly*

Low median—*creating jobs, education, housing shortages, raising life expectancy*

Possible solution will vary.

Enrichment/Extension

1. Create a bulletin board display showing characteristics of high and low median age nations and color in the two categories on a world map.
2. Create graphs or charts comparing the statistics from this lesson.
3. Bring in magazine ads that reflect advertisers awareness of the median age in the United States, and identify how some product advertisements are aimed at different age groups of our society.
4. Write or present an oral report (or engage in a debate/panel discussion) on this question: Should the age of political leadership be similar to the median age of a nation? Why or why not?

Name________________
Date________________

Median Ages of Nations

Part A.

Study the chart below before completing part A. Following the chart, several nations are listed, accompanied by one statistic. Using the information that you learned, identify each nation as either high median age or low median age. Be able to explain your choice. Complete part A of the handout as directed.

High median age nations	**Low median age nations**
High life expectancy	Low life expectancy
Low birthrate (less then 2%)	High birthrate (more than 3%)
Low death rate	High death rate
"Newer" nation or culture	"Older" nation or culture
Temperate climate	Tropical climate
"Northern" regions of earth	"Southern" regions of earth
Rich nation	Poor nation
Economically developed	Economically underdeveloped
Industrial nation	Agricultural nation
High per capita income	Low per capita income
Low percentage of population under age of fifteen years old	High percentage of population under age of fifteen years old
Low annual population growth	High annual population growth
High literacy rate	Low literacy rate

Name________________
Date________________

1. Complete the chart by using information given in the preceeding chart.

   Example: Sweden—21 percent of population less than 15 years. High median age.

Nation	Statistic	Median age of nation
a. Pakistan	Annual population growth rate—3.0%	
b. Burundi	Life expectancy for males—38 years	
c. Greece	Annual population growth rate—.1%	
d. Albania	Percentage of population in agriculture—47%	
e. Bangladesh	Percentage of population over 65 years—3%	
f. Zimbabwe	Percentage of population under 15 years—48%	
g. Niger	Percentage of population in agriculture—90%	
h. Sweden	Percentage of population over 65 years—18%	
i. Norway	Life expectancy for males—74 years	
j. Luxembourg	Percentage of population under 15 years—17%	

2. Listed below is a nation with one statistic. Classify the nation as either high or low in each of the related characteristics designated below.

   The life expectancy of a person living in the African republic of Chad is forty-one years old.

   a. Literacy rate

   b. Median age

   c. Percent of population under 15 years

   d. Percent of population in agriculture

   e. Percent of population living in cities

3. What other characteristics might you expect to find in Chad?

Name________________
Date________________

Part B.

Use the chart as a resource in completing the activity that follows.

Nation	Percentage of population less than 15 years	Life expectancy M	Life expectancy F	Annual population growth rate	Percentage of population in agriculture	Literacy rate
High						
United States	22	72	79	.7	2.5	97
Canada	21	72	82	.7	4.0	99
United Kingdom	19	74	80	.3	1.0	99
Japan	17	76	82	.3	7	99
Israel	31	76	80	1.4	6	92
Australia	22	74	81	.7	6	99
Low						
Brazil	35	57	67	1.3	31	81
Panama	35	72	78	2.0	27	87
Mexico	38	69	77	2.2	28	90
Egypt	40	59	63	2.3	34	44
India	36	58	59	1.8	67	48
China	28	67	69	1.1	60	70
Vietnam	39	63	68	1.9	65	88

Assume that you are a cabinet member of the government of one of the nations in the world. Your country has a ________________ median age. What you must do as a cabinet is to identify the problems and concerns that your nation will face in the next twenty years. If possible, try to suggest how you can go about solving the problems so that you can eliminate them in the future.

Problems and concerns

Possible solutions

Lesson 39
Cultural Diffusion—Ideas to Go

Objectives

- To identify the geographic theme of movement in the transfer of ideas and cultural features
- To identify how new ideas result in cultural changes

Notes to the Teacher

Students examine geography from the perspective of the movement of goods and ideas and how that movement affects cultures. Students need to understand that as people move within regions and among various countries, they take their culture with them—the way they dress, how they build their homes, the kind of food they eat, the language they speak, the religions they practice, the kind of governments they support, the music and recreation they enjoy, etc. When new ideas enter an area, they are often adopted by the people who live in that area. Eventually, their traditional culture changes. Sometimes this happens through friendly trade contacts and sometimes this happens through military contact and/or occupying troops. Today, it happens via television and movies. Satellite dishes make every country in the world aware of the latest dances in America, the latest fashions in Paris, the latest Japanese automobiles, fads, etc. Ideas move with electric speed and change traditional cultures by blending western ideas into eastern cultures or vice versa. Our trade goods change cultures as well. It is not difficult to get electronic goods from the east at prices so low that almost anyone can afford them—changing cultures yet again.

Students read a play and identify the various goods and ideas that moved among countries. They note these changes, chart them, and explain how the new ideas/goods got to these places. Students color a world map and key it to these countries and indicate the ideas that were exchanged.

Procedure

1. Bring in several items that are not a part of the traditional U.S. culture, such as chopsticks, a Koran, a Mexican sombrero, etc. Place several of these items before the class and ask them to name the country they think of when they see these things. Then ask them if they can be found anywhere in the United States. Allow them a few minutes to suggest where they might be found. Many cities as well as small towns have Oriental restaurants where chopsticks could be found.
2. Explain that when we send goods to other countries and they send things to us we are trading, but we are also exchanging cultural information about one another. We do the same thing with ideas when we exchange them. Often we get so used to seeing and using these things that they become a part of our own culture. Sometimes after a war, a country sends its troops temporarily into another country. These troops bring ideas and things from home to make themselves more comfortable. The people living in those countries often start using these ideas and goods and enjoy them just as the soldiers do. Even language is changed in this way.
3. Assign roles to students and have students read the play, **Handout 90**.
4. Identify all the countries that are mentioned in the play and what was exchanged—where it came from and where it went. Discuss the circumstances that introduced the new ideas—war, peace, trade, etc.? Construct a chart on the chalkboard to include all of this information.

Suggested Responses:

Country of origin	Exchange/item	Receiving country	How/why did it happen?
Britain	Soccer	United States, France, Germany, Nigeria	Introduced through immigration
United States	Baseball	Japan	Postwar occupation
United States	Country western music	Japan	Postwar occupation
West African Countries	Blues and jazz music	United States	Descendants of African slaves
North African Countries	Moorish language, rice, saffron, cumin, anise, almonds, bananas	Spain	Moorish occupation of Spain
United States	Euro-Disney, french fries	France	Economic invasion
Mexico	Chocolate	Spain	Exploration and occupation in fifteenth century
South America	White potatoes	Spain	Exploration and occupation in fifteenth century
North America	Tomatoes	Spain	Exploration and occupation in fifteenth century
Italy	Pizza, French and Spanish language (from Latin, musical terms)	United States, France, Spain	Occupation/immigration
Germany	English language	Great Britain	Occupation
Britain	High tea	United States	English tradition/heritage
China	Tea	Great Britain	Occupation/trade
Kenya	Coffee	Great Britain	Occupation
Great Britain	English language and English-style of government	Nigeria	Occupation/trade

5. Provide each student with a world map using **Handout 91**. Tell them to make a key that is color coded. Have students select a different color for each country in the play and color the countries. Write the product, idea, language, etc., on or near the country that received it, and draw arrows to the country of origin. Divide the class into small groups and have them compare and discuss their completed maps.

6. Discuss what students have discovered about *cultural diffusion*. Introduce the term *cultural diffusion* and explain that this happens because of movement of ideas and goods among the various peoples of the world. This movement has increased greatly today because of our greater technological capabilities, which reduce the influence of distance and other physical barriers that used to impede this movement of ideas and goods.

Enrichment/Extension

1. Explore your community for restaurants that represent various ethnic groups. Collect sample menus and compare the kinds of foods prepared. Are there commonalities? Share your findings with classmates.

2. Make a recipe book of recipes from around the world and design a cover. Draw maps of each country where you have collected recipes. Share your recipe book with your families and others.

3. Research a set of ten countries and discover their most popular sport. Make a graph/chart/poster illustrating your findings.

Name________________
Date________________

On the Internet

Read the dialogue exchange among these students on the Internet. Note the different nations they represent and the various cultural features that they share.

Narrator: It is late in the small Ohio town, almost 11 p.m. Everyone in the house was asleep, but Steven had a homework assignment to do on the Internet. He was even looking forward to it! Hard to believe, right? Well, he was excited because all of his penpals were going to be on-line and they were going to talk to each other by computer. Some people were up at even stranger hours than he was.

Steven began the sequence of entries to gain access to the Internet. Impatiently he waited for the modem to function and the computer to begin the process that would connect them all at the same time. Everyone could be in the conversation with Steven. He was thinking about what he wanted to say. Should he tell them that he lived close to Lake Erie, a lake named for a Native American group who once lived there? That was just one of the many Native American names of cities and foods that he could think of. Maybe they wouldn't be interested. Oh well, he thought, I'll see what happens.

Logging in on the Internet were Steven in the U.S.A., Yoki in Japan, Reggie in Italy, Sarah in Great Britain, Iboga in Nigeria, Franco in Spain, Anna in Germany, and Jean in France.

Steven: Hurry up, computer. I can't stand the wait!!!......(Finally, the screen began to flash the various codes that said he was connected.) Hello, ... Hello, this is Steven. I am calling tonight to talk to you about things that we have in the United States that came here from other countries or things that we have sent to other countries. My teacher said that many people have come to the United States from all over the world bringing with them their cultures. They shared those things with us and they learned to do things like U.S. citizens. I'm trying to find out about some of those cultural trades tonight. What are your favorite sports?

Sarah (Great Britain): Soccer! Hello, Steven. The British started the game of soccer. People here are crazy about it. We call it football.

Reggie (Italy): Football!! My country knows the game well and loves it. It came to us from Great Britain many years ago. Do you play, Steven?

Steven: Yes, I really like to play soccer. We have some professional teams, but more people in the United States like football and baseball.

Yoki (Japan): Baseball!! I love to go to baseball games. My friends really enjoy the game. Many Japanese have become very western in this way. You Americans brought the game to Japan after World War II. Almost everyone fell in love with it. By the way, Americans brought country-western music over here, too. I really like to "boot scoot" my way around the dance floor at school dances.

Steven: Oh, Yoki. I'm glad you enjoy that part of our culture, but country western isn't really my thing. I like rock and rap. We can thank African Americans for helping the United States musically. They brought their talents from their African countries and passed them on, and now we have the blues and jazz.

Franco (Spain): Oh, yes. Africans influenced my country, too. The Moors invaded us several hundred years ago and over 4,000 words from their language became a part of our language. They also introduced rice to us and wonderful spices like saffron, cumin, and anise as well as almonds and bananas. I like to eat, can you tell?

Name_______________
Date_______________

Jean (France): In France, we like to keep our language "pure." There are even laws that say we must use French unless there is no French word for the item. So many English-speaking businesses and tourists come to France, it is hard not to find some English words being used. Did you know that Disney has a park here?

Steven: Yes, I know. That's another American idea that has crossed the Atlantic to invade Europe. I have been to the one in Orlando, Florida. I had a great time in Epcot visiting all the different pavilions for the various countries around the world. Have you gone to Disneyworld in France, Jean?

Jean (France): Yes, my grandparents took me on my birthday. I really like Mickey Mouse, but some of it was—well, just too American—sorry, I do not mean to hurt your feelings.

Anna (Germany): My parents took me to France to the Disney park at Christmas. I had a really good time. I really like the American food they served—like French fries.

Jean (France): I like the French fries, too, even if they were an American invention.

Franco (Spain): Food—Food, Do you like hot chocolate on a cool day? Did you know the Spanish found chocolate in Mexico when they conquered the Aztec Indians in 1522? The Aztecs liked their chocolate without sugar and milk, but when it came to Spain, we added sugar and milk and made it sweet and to our taste. Most people drink it our way. Speaking of food, we also brought white potatoes from the Incas in South America, and we found tomatoes in North America. All of those things go into good Spanish cooking!

Steven: Gee, Franco, you do like to eat! How about you Reggie?

Reggie (Italy): Si, I like to eat. You know, down in Napoli, they developed the first pizza—oh!!! They are wonderful! In Italy, we have many wonderful things that we got when we contacted other countries, many that were part of our empire. Sometimes we learned from them and sometimes they learned from us. Latin was spoken in many places and other languages grew out of it like French and Spanish. Many words today were taken from Italy, especially military terms like corporal, captain, colonel, general, and artillery. Italians loved music and many musical terms came from us, too, like soprano, piano, tempo, libretto, etc. After World War II, while the American were here we picked up some American words that have become part of our language—words like popcorn, shopping, weekend. So many!! I can't remember them all.

Steven: That's great! I know all those words. I could almost speak another language. What do you all think?

Anna (Germany): Well, German might be easy for you, too. We have many words that would be familiar because the English language developed from the German language brought to Great Britain when German tribes controlled that country.

Steven: I didn't know that! That's pretty amazing. What do you think about that Sarah?

Sarah (Great Britain): I did know about that. We study about those times in school. By the way do you know anything about tea?

Steven: A little. My mom likes it. Sometimes she goes with my Aunt Sue to a hotel restaurant in the city for a tea party in late afternoon. She says they serve good little sandwiches and pastries.

Name________________
Date________________

Sarah (Great Britain): That sounds like a real British "high tea." I had heard that some Americans were doing that. We British have done so for many years. It's a lot like your dinner time if it's a full tea with lots to eat and plenty of tea.

Yoki (Japan): Well, the British didn't know about tea until they began to trade and to control some of the Asian countries. The Chinese had the first tea and it was not legal to take tea plants out of the country, but eventually some were taken out and grown in India and other southeastern Asian countries. British traders took it home and it really caught on. Japan traded with China in the early days. We Japanese even have a special tea ceremony when we serve tea.

Iboga (Nigeria): Finally, I am in! My computer was not working. Sounds like I missed some, but I was listening to the "tea talk," and it reminded me of a friend in Kenya. They have coffee plantations there and much of the coffee once went to Britain as well as to other countries. Britain used to control them as a colony as they did my country. That is why English is spoken here. Our educational system and governments are run very much like the British system, too. They caused many changes in our culture.

Steven: Hello, Iboga. Do you play soccer? I heard that people really like it there?

Iboga (Nigeria): Oh, yes. We play at school all the time, and we have some good professional teams.

Steven: Our time is nearly up everyone. It was great to talk to you. I have learned so much. Please keep writing to me and let me know what you are doing.

All the Students: Good-Bye, Steven. Good luck on your report!!!

Name__________________
Date__________________

Points of Origin

Make a key for your map by choosing a different color for each country you read about in this lesson. Color each of the countries to match the key. Write the product, idea, language, etc., around the map and draw a line connecting the product or idea to its country of origin.

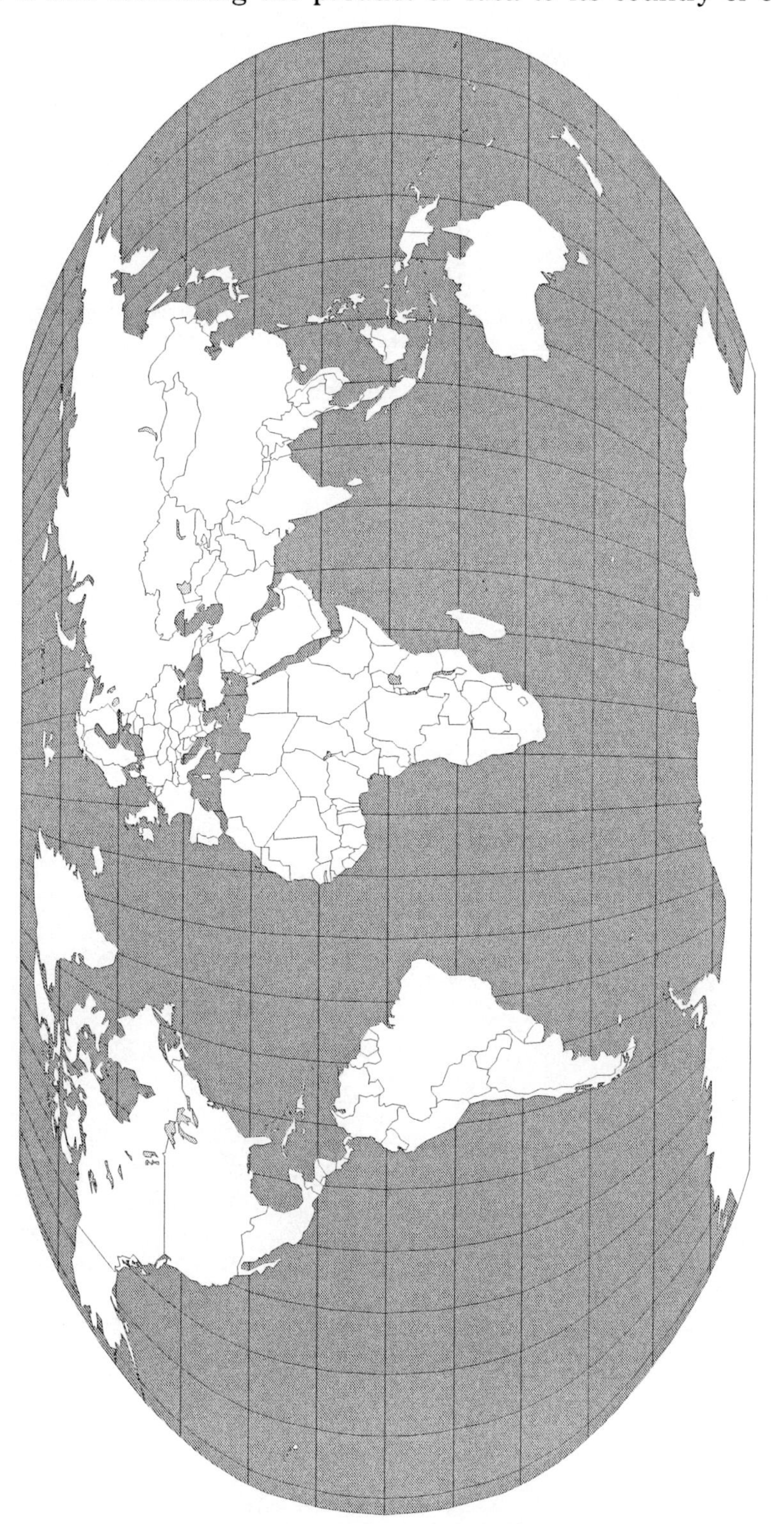

Lesson 40
Subcultures—A Study in Contrasts

Objectives

- To understand the concept of subcultures
- To contrast Native American and Maori subcultures

Notes to the Teacher

Often there exists within the dominant culture of a nation a minority group. Sometimes, over a long period, the minority group takes on the characteristics and traits of the dominant culture so that it becomes nearly indistinguishable from the dominant culture, in which case the minority group is fully integrated, incorporated, and assimilated into the dominant culture. However, when this does not happen and the minority culture retains its own traits, characteristics, and norms, it is defined as a subculture. At times, as in the case of many Native Americans, the majority government restricts the subculture to reservations, or lands set aside in specific locations, where the subculture is expected to live.

Often a major difficulty encountered by subcultures derives from the ethnocentric attitudes of the dominant culture. Usually, the subculture wishes merely to be accepted on its own terms. However, the majority may expect the subculture to accept the majority's ways, and to be incorporated, assimilated, and integrated into its society. If this does not happen, the majority culture may treat the subculture with disdain or prejudice. The majority culture may force the subculture onto reservations or tribal homelands, or worse still, strive for the subculture's extinction (genocide). Too often the majority views subcultures as tourist attractions or case studies for anthropologists intent on recording the culture, arts, and folklore of the subculture before it becomes extinct, rather than show concern for the preservation of the subculture itself.

Students read about the experiences of the Maori of New Zealand and the Native Americans of the United States. Students prepare a chart which compares the two subcultures and why the Maori have been more successful than Native Americans at social integration and cultural preservation.

Procedure

1. Begin by reviewing the concept of culture. Introduce the concepts of majority or dominant cultures and minority or subcultures using material in the Notes to the Teacher. Point out the frequency of ethnocentric attitudes of the majority and manifestations they may take.
2. Locate New Zealand, home of the Maori, on a wall map, and note that Native American reservations of vastly different sizes are scattered in many states of this country. Distribute **Handout 92**, parts A and B. Allow students time to read the two selections and to reflect on similarities and differences between the two subcultural groups. Compile a list of similarities and differences and place the listing on the chalkboard.
3. Distribute **Handout 93**. The completion of the comparative chart can be directed as either an individual, paired, or small-group activity. Ask students to volunteer to share their answers with the class. Use students' responses as a springboard for discussion.

Suggested Responses:

	Native Americans	Maori
Social equality	Often treated as second-class citizens	Treated as equals in society
Government participation	No direct representation	Four members in Parliament
Land holding	Government controlled reservations	Independent, communally-owned land holdings
Social incorporation	Often isolated on reservations; many in isolated pockets in cities	Own and live on communal land; many integrated in cities
Occupations	High unemployment on reservations	Integrated into work force
Medical help provided	Yes, on reservations; not always adequate	Free anywhere in the country
Education provided	Limited access on reservations	Free anywhere in the country
Other		

4. For a culminating discussion, ask the following question: How or why have the Maori of New Zealand been more successful at incorporation and subcultural preservation than the Native American? (*Treaties with England successful; unification under a single leader; involved directly in the government; more direct control of their land.*)

Enrichment/Extension

1. Prepare a report (two to three minute oral, one to two page written) on one of the following subcultural groups:
    - Aborigines of Australia
    - Guaymi tribe in Panama
    - Mosquita tribe in Nicaragua
    - Hopi tribe in Arizona
    - Navajo in Western United States
    - Tamil in Sri Lanka
    - Sikh in India
    - Tribal Homelands (Lesothos and Swaziland) in South Africa
    - Basques in Spain and France
    - Yaqui in Mexico
    - Gypsies in Western Europe
    - Ethiopian Jews in Israel
    - Beothuks in Newfoundland
2. Prepare a report (two to three minute oral, one to two page written) on a person from either the Native American or Maori subcultures—Russell Means, Kiri Te Kanawa, etc.

Name________________
Date________________

The Maori of New Zealand

Part A.

Read the following selections. Take note of similarities and differences. Be prepared for class discussion.

In the 1300s, the first wave of people arrived in New Zealand: the Maori. Coming from Tahiti in the Pacific, the Maori were the first to settle the uninhabited islands. Tribalistic, artistic, and territorial, they established a culture based upon hill forts ruled by a chief, advised by his priest, and populated by his family tribe.

In 1769, when the first Europeans began to arrive, there were 200,000 Maori living in New Zealand. Fearing that the lawlessness of passing sailors and greedy colonists would destroy them, the Maori petitioned King William IV of England in 1831 for British protection. The Treaty of Waitangi, signed in 1840, granted tribally-held lands to the Maori. However, by 1858, colonists outnumbered the Maori. In 1860, in order to unify themselves against what they now recognized as a majority population, the Maori tribes convened and elected a Maori king to speak for the whole Maori culture. Despite brief "Maori Wars" in the 1860s, the Maori have been incorporated, but not assimilated, into the culture of modern New Zealand.

Today, they represent 8 percent of the population of New Zealand and have a greater growth rate than the majority culture. As a result of the Treaty of Waitangi, they own communally four million acres of tribal lands. The Maori enjoy equal rights and participation in government and in society. Four members of Parliament must be Maori; there is a Minister of Maori Affairs, a Maori Educational Foundation, and a Maori Council of elected tribal executives. A separate Maori Land Court settles on-going land disputes. They are entitled to free education, medicine, and medical and dental services, like the majority, until age sixteen. Many Maori hold jobs in the civil service, farm, serve in the military, or work in a variety of trades or professions.

Name_______________
Date_______________

Native Americans of the United States

Part B.

Most of this country's Native American population of 875,000 live on reservations of varying sizes. The Navajo reservation, which covers parts of Arizona, New Mexico, and Utah, is home to nearly 96,000 members, while the Western Pequot reservation in Connecticut, one of the tiniest reservations, has 16 residents. Viewed historically as separate nations, these reservations are more like domestically dependent nations, existing under the control of the United States Congress, making the tribal governments more like county governments than state governments.

Although some reservations are economically sound, due to tourism or mineral, oil, or water rights profits, most of the reservations suffer economically. If located in isolated areas, there may be poor access to educational facilities, jobs, or markets for goods. Too often the land will not produce enough food for agricultural self-sufficiency or herding. When the young people leave the reservations to earn a decent wage in the city, the reservations are left with women, children, and the elderly. Usually, when a reservation has government facilities, the women are those who find clerical work. Further, the traditional Indian roles of men as warriors and hunters, and women as farmers, are not applicable to modern economic survival. When many of these problems occur at the same time, the reservations, these separate nations, seem like poor, underdeveloped countries.

Although life expectancy for Native Americans has risen dramatically in recent years, it is still more than three years lower than that for white Americans. Other problems are more serious. Unemployment is 13 percent for Native Americans as a whole, but figures for some large reservations are nearly three times that much. While 9.8 percent of white Americans live below the poverty line, the figure for Native Americans is 30.9 percent. The situation is worse for children where nearly 38 percent of Native American children live in poverty. The most recent census figures put the number of Native Americans having no indoor toilets at 20 percent while 50 percent have no phone and 16 percent live without electric lights. Median annual income for Native Americans is $20,025 compared to $31,435 for white Americans. Many struggle to survive making native crafts for tourists. Nearly 12 percent of Native American women have no prenatal care, and birth defects among Native Americans are nearly 50 percent more frequent than among Hispanic Americans. Alcoholism and related diseases account for more than 20 percent of deaths among Native Americans; in fact alcohol-related deaths are more than five times as prevalent among Native Americans as among white Americans. Although the government funds hospitals and clinics on reservations, the unavailability of trained doctors willing to work for lower incomes on reservations has often led to poor medical service and even caused the closing of hospitals on reservations. Education on reservations, too, lags behind that offered to other Americans, in part because of the unavailability of qualified teachers willing to work for lower salaries on reservations.

Name________________
Date________________

Comparing Subcultures—Native American and Maori

After reading the passages on Native Americans in the United States and the Maori in New Zealand, complete the chart below comparing the two groups according to the areas indicated.

	Native Americans	Maori
Social equality		
Government participation		
Land holding		
Social incorporation		
Occupations		
Medical help provided		
Education provided		
Other		

Acknowledgments

For permission to reprint all works in this volume, grateful acknowledgment is made to the following holders of copyright, publisher, or representatives.

Lesson 3, Handout 7

From *How to Use Maps and Globes* by Helen Carey. Copyright © 1983. Reprinted by permission of Franklin Watts.

Lesson 3, Handout 8

E.N. Askov and K. Kamm, from *Study Skills in the Content Area*. Copyright © 1982 by Allyn and Bacon. Reprinted/adapted by permission.

Lesson 13, Handout 35

Excerpt from *Amazon Up Close* by Pamela Bloom. Copyright © 1997, Hunter Publishing.

Lesson 31, Handout 80

From *African Folktales* by Roger D. Abrahams. Copyright © 1983 by Roger D. Abrahams. Reprinted by permission of Pantheon Books, a division of Random House.